Practice Papers for SQA Exams

Standard Grade | Credit

Physics

Text © 2009 Martin Cunningham
Design and layout © 2009 Leckie & Leckie

01/150609

ISBN 978-1-84372-775-0

Published by
Leckie & Leckie Ltd, 4 Queen Street, Edinburgh, EH2 1JE
Tel: 0131 220 6831 Fax: 0131 225 9987
enquiries@leckieandleckie.co.uk   www.leckieandleckie.co.uk

A CIP Catalogue record for this book is available from the British Library.

Leckie & Leckie Ltd is a division of Huveaux plc.

Questions and answers in this book do not emanate from SQA. All of our entirely new and original Practice Papers have been written by experienced authors working directly for the publisher.

Introduction

Layout of the Book

This book contains practice exam papers, which mirror the actual SQA exam as much as possible. The layout, paper colour and question level are all similar to the actual exam that you will sit, so that you are familiar with what the exam paper will look like.

The answer section is at the back of the book. Each answer contains a worked out answer or solution so that you can see how the right answer has been arrived at. The answers also include practical hints on how to tackle certain types of questions, details of how marks are awarded and advice on just what the examiners will be looking for.

Revision advice is provided in this introductory section of the book, so please read on!

How To Use This Book

The Practice Papers can be used in two main ways:

1. You can complete an entire practice paper as preparation for the final exam. If you would like to use the book in this way, you can either complete the practice paper under exam style conditions by setting yourself a time for each paper and answering it as well as possible without using any references or notes. Alternatively, you can answer the practice paper questions as a revision exercise, using your notes to produce a model answer. Your teacher may mark these for you.

2. You can use the Topic Index to find all the questions within the book that deal with a specific topic. This allows you to focus specifically on areas that you particularly want to revise or, if you are mid way through your course, it lets you practise answering exam style questions for just those topics that you have studied.

Revision advice

Work out a revision timetable for each week's work in advance – remember to cover all of your subjects and to leave time for homework and breaks. For example:

Day	6pm–6·45pm	7pm–8pm	8·15pm–9pm	9·15pm–10pm
Monday	Homework	Homework	Business Management revision	Chemistry revision
Tuesday	Maths revision	Physics revision	Homework	Free
Wednesday	Business Management revision	Modern Studies revision	English revision	French revision
Thursday	Homework	Maths revision	Chemistry revision	Free
Friday	Geography revision	French revision	Free	Free
Saturday	Free	Free	Free	Free
Sunday	Modern Studies revision	Maths revision	Physics revision	Homework

Make sure that you have at least one evening free a week to relax, socialise and re-charge your batteries. It also gives your brain a chance to process the information that you have been feeding it all week.

Arrange your study time with a break between sessions. Try to start studying as early as possible in the evening when your brain is still alert and be aware that the longer you put off starting, the harder it will be to start!

Study a different subject in each session, except for the day before an exam.

Do something different during your breaks between study sessions – have a cup of tea, or listen to some music. Don't let your 15 minutes expand into 20 or 25 minutes though!

Have your class notes and any textbooks available for your revision to hand as well as plenty of blank paper, a pen, pencil, rubber, etc. You may like to make keyword sheets like the following physics example:

Keyword	Meaning
Half-life	Time taken for the activity of a radioactive source to half.
Ultrasound	Sound vibrations with a frequency greater than 20 000 Hz (i.e. above the range of human hearing).
Wavelength	The distance between two identical neighbouring parts of a wave, e.g. the distance between two neighbouring wave crests.

Finally, forget or ignore all or some of the advice in this section if you are happy with your present way of studying. Everyone revises differently, so find a way that works for you!

Transfer Your Knowledge

As well as using your class notes and textbooks to revise, these practice papers will also be a useful revision tool as they will help you to get used to answering exam style questions. You may find as you work through the questions that they refer to a situation or an example that you haven't come across before. Don't worry! You should be able to transfer your course knowledge to a new example. The enhanced answer section at the back will demonstrate how to apply your course knowledge in order to answer the question successfully.

Command Words

In the practice papers and in the exam itself, a number of command words will be used in the questions. These command words are used to show you how you should answer a question – some words indicate that you should write more than others. If you familiarise yourself with these command words, it will help you to structure your answers more effectively.

Command Word	Meaning/Explanation
Name, state, identify, list	Giving a list is acceptable here – as a general rule you will get one mark for each point you give.
Suggest	Give more than a list – perhaps a proposal or an idea.
Outline	Give a brief description or overview of what you are talking about.
Describe	Give more detail than you would in an outline, and use examples where you can.
Explain	Discuss why an action has been taken or an outcome reached – what are the reasons and/or processes behind it.
Justify	Give reasons for your answer, stating why you have taken an action or reached a particular conclusion.
Define	Give the meaning of the term.
Compare	Give the key features of 2 different items or ideas and discuss their similarities and/or their differences.
Predict	Work out what will happen.

In the Exam

- If there is something which you are liable to forget, as soon as the exam starts, write down brief notes about it on the back of the question paper – do this lightly in pencil.
- Watch your time and pace yourself carefully. Work out roughly how much time you can spend on each answer and try to stick to this.
- Be clear before the exam what the instructions are likely to be. The practice papers will help you to become familiar with the exam's instructions.

- You may attempt the questions in any order – many candidates starts with their 'favourite unit' questions and work through to the questions covering their 'least favourite unit'.

- Read each question thoroughly before you begin to answer it – make sure you know exactly what the question is asking you to do.

Ask yourself:

- What am I being asked?

- Do I need to use a relationship/formula? If so, check the Physics data book

Plan your descriptive answer by jotting down keywords, a mind map or reminders of the important things to include in your answer. Cross them off as you deal with them and check them before you move on to the next question to make sure that you haven't forgotten anything. you could do this lightly in pencil on the exam paper.

When performing calculations:

- Check the relationship/formula you are using in the Physics data book – always write down the formula.

- Show every step in the calculation – don't be tempted to miss out steps!

- Always include a unit with your final answer (unless you are being asked to calculate *gain*, which does not have a unit).

If you can't work out the numerical answer to the first part of a question and you need to use this number for the next part, make up any number and use it in the second calculation – you will be awarded full marks for the second part if you perform the calculation correctly.

Don't repeat yourself as you will not get any more marks for saying the same thing twice. This also applies to annotated diagrams which will not get you any extra marks if the information is repeated in the written part of your answer.

Give proper explanations. A common error is to give descriptions rather than explanations. If you are asked to explain something, you should be giving reasons. Check your answer to an 'explain' question and make sure that you have used plenty of linking words and phrases such as 'because', 'this means that', 'therefore', 'so', 'so that', 'due to', 'since' and 'the reason is'.

Use the resources provided. Some questions will ask you to 'describe and explain' and provide an example or details of an experiment for you to work from. Make sure that you take any relevant data from these resources.

Good luck!

Topic Index

Topic	Exam A	Exam B	Exam C	Exam D	Questions I found difficult/need help with
Telecommunication	1 2	1 2 3 7 (a)	1 2 3 15 (b)	1 2	5
Using Electricity	3 (a) 3 (b) 3 (c) 4	4 5	4 5 13 (d)	3 4	2
Health Physics	5 6 7	6 7 (b) 15 (a)(iii)	6 7	5 6	4
Electronics	3 (d) 8 9	8 9 12 (c)	8 9	7 8	3
Transport	10 11	10 11	10 11	9 10 14 (a)	1
Energy Matters	12 13	12 (a) 12 (b) 13	12 13 (a) 13 (b) 13 (c)	11 12 14 (b) 14 (c)	
Space Physics	14 15	14 15 (a) (i) 15 (a) (ii) (b)	14 15 (a)	13 14	

Credit Physics

Practice Papers for SQA Exams	Time allowed: 1 hours, 45 minutes	**Exam A**

Fill in these boxes and read what is printed below.

Full name of school

Town

Forename(s)

Surname

Read each question carefully.

Attempt **all** questions. You may answer them in any order.

Write your answers in the spaces provided.

Write as neatly as possible. If you decide to change your answer, cross it out neatly before rewriting it.

Answer in sentences wherever possible.

Any data needed will be on the data sheet at the back of the book.

Use appropriate numbers of significant figures in final answer.

Leckie × Leckie

Scotland's leading educational publishers

Marks

KU	PS

1. The diagram represents a picture being formed on the screen of a black and white television set:

special phosphor
coating inside screen
which glows when
hit by electrons

electron beam

electron gun
fires electrons at
screen

magnetic coils
move electron beam
across and down screen.

Each picture is built up of 625 lines. Each second, 25 separate pictures appear on the screen.

(a) Calculate how many lines are built up on the screen in 1 second.

> *Space for working and answer*

1

(b) Why do our eyes/brain believe there is a continuous picture on the screen?

....

....

....

2

(c) By altering the instructions sent to the electron gun, any point on the screen can be made brighter or darker.

Explain how the electron gun is used to **increase** the brightness at a particular point on the screen.

....

....

1

2. (a) (i) What term is used to describe the ability of waves to travel around large objects, such as hills?

....

1

(ii) A house is situated behind the large hill.

Complete the two diagrams below (by drawing the path of the waves on the right hand side of the hill) to show whether television or radio reception will be better at the house.

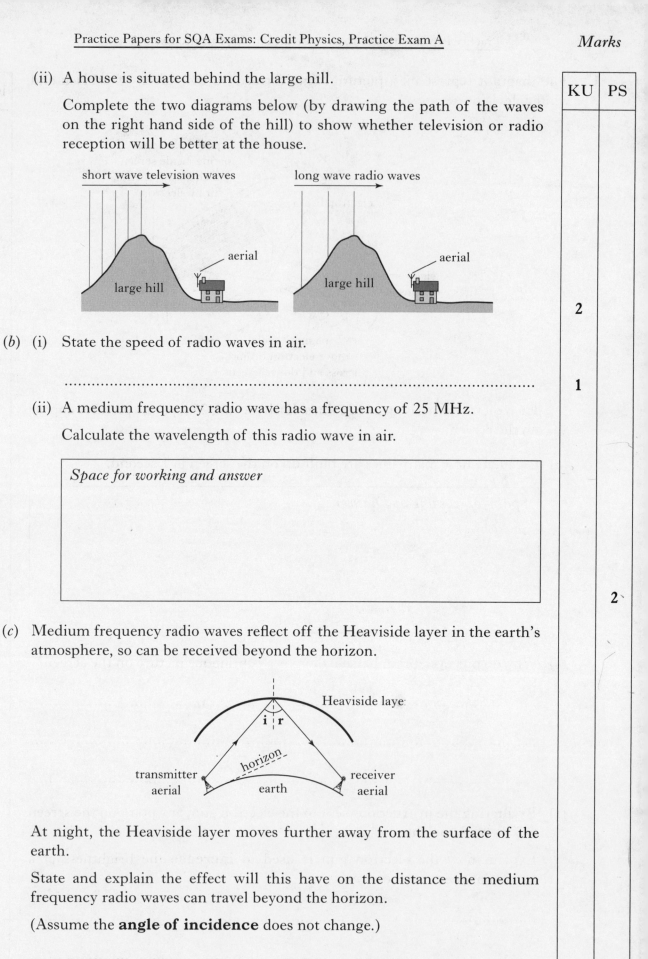

(b) (i) State the speed of radio waves in air.

....

(ii) A medium frequency radio wave has a frequency of 25 MHz.

Calculate the wavelength of this radio wave in air.

Space for working and answer

(c) Medium frequency radio waves reflect off the Heaviside layer in the earth's atmosphere, so can be received beyond the horizon.

At night, the Heaviside layer moves further away from the surface of the earth.

State and explain the effect will this have on the distance the medium frequency radio waves can travel beyond the horizon.

(Assume the **angle of incidence** does not change.)

....

....

....

KU	PS
2	
1	
	2
	2

Marks

KU	PS

3. A toy robot contains a 12 V battery, a small electric motor to make it move across the floor and two lamps for its eyes.

The circuit diagram for the robot is shown.

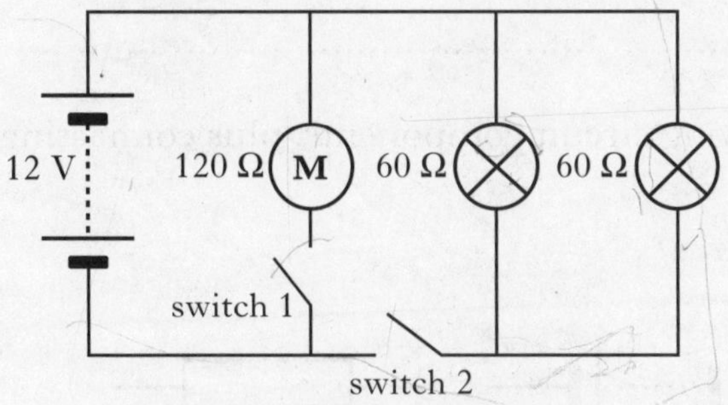

(a) Each lamp has a resistance of 60 Ω.

Calculate the power rating of each lamp.

Space for working and answer

2

(b) Both switches are closed, turning on the electric motor and both lamps.

The electric motor has a resistance of 120 Ω.

Calculate the total resistance of the electric motor and both lamps in the circuit.

Space for working and answer

2

(c) When the electric motor and both lamps are switched on, the 12 V battery provides a steady current of 500 mA.

Calculate the total charge which will flow through the battery when the electric motor and both lamps are switched on for 3 minutes.

Space for working and answer

KU	PS

2

(d) A toy designer suggests the toy robot could be improved by replacing the lamps with light emitting diodes.

  (i) Suggest one advantage a light emitting diode would have over a lamp.

  ....

  (ii) The toy designer has the following circuit components, plus connecting wire:

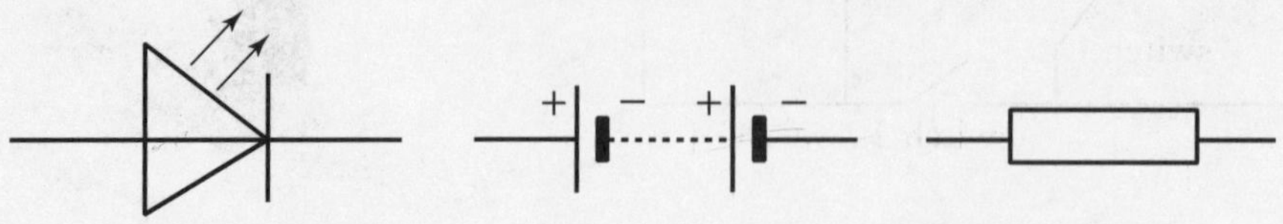

  Draw a circuit diagram to show how the toy designer could connect these circuit components to allow the light emitting diode to light.

  Space for answer

4. The diagram shows the wiring for an electric heater with a **metal case**:

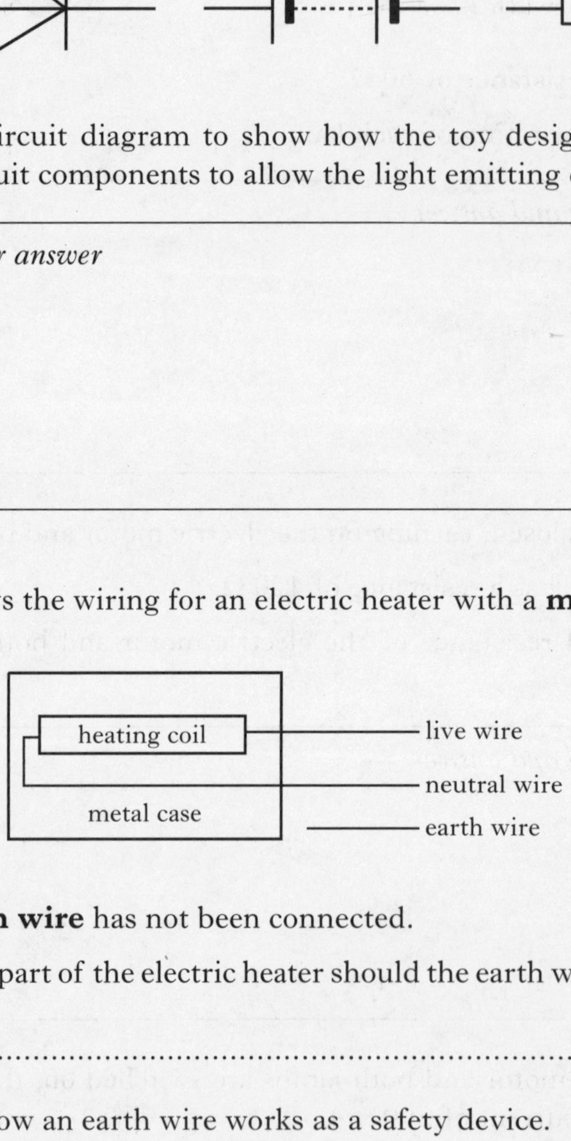

  (a) (i) The **earth wire** has not been connected.

      To which part of the electric heater should the earth wire be connected?

  ....

  (ii) Explain how an earth wire works as a safety device.

  ....

  ....

  ....

Marks

(iii) A fuse and switch are connected in the live wire.

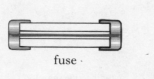

fuse

switch

Explain why the fuse and switch must **not** be connected in the **neutral** wire.

....

....

1

(b) The diagram shows two household circuits, a ring main circuit and a lighting circuit, connected to a consumer unit.

Both circuits have an earth connection – not shown on the diagram.

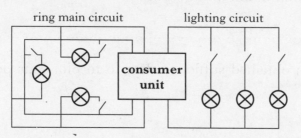

(i) In modern consumer units, metal wire fuses have been replaced by miniature circuit breakers.

State **one** reason why miniature circuit breakers are now used in preference to fuses.

....

1

(ii) State and explain **one** difference between a ring main circuit and a lighting circuit.

....

....

....

2

(iii) Electric cookers must never be connected to the ring main circuit. They require a separate circuit.

Explain why.

....

....

....

1

KU | PS

Marks

KU	PS

5. (*a*) A school technician sets up the apparatus below to measure the half-life of a radioactive sample which emits alpha particles.

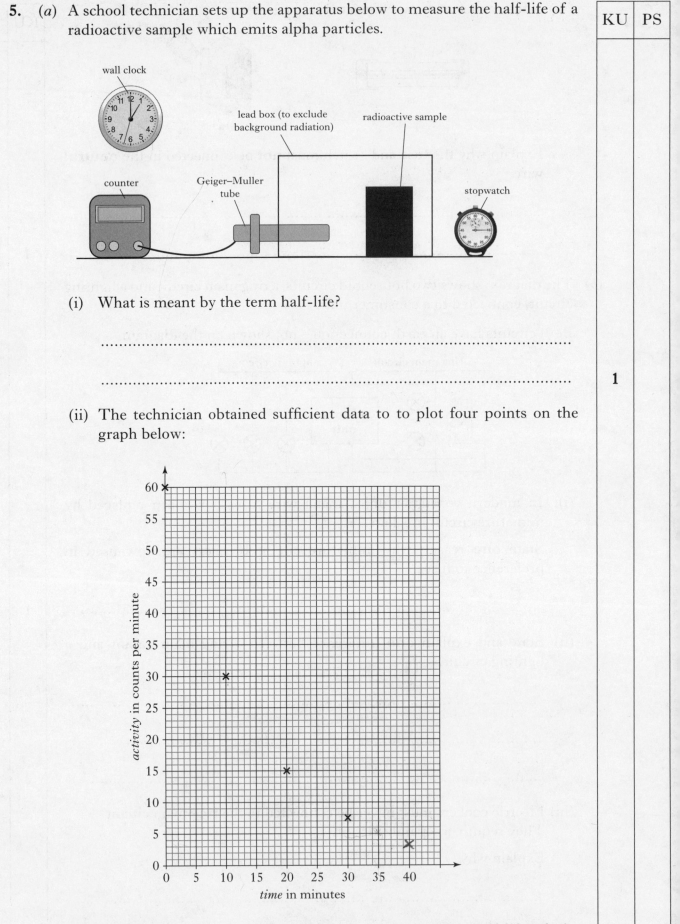

(i) What is meant by the term half-life?

....

....

1

(ii) The technician obtained sufficient data to to plot four points on the graph below:

Use the points plotted on the graph to determine the half-life of the radioactive sample.

	KU	PS

Space for working and answer

2

(iii) Use the graph to estimate the value for the activity of the radioactive substance **35 minutes** after measurement started.

....

1

(iv) Describe how the technician could have obtained her data.

....

....

....

2

(b) Alpha particles ionise atoms.

(i) What is **ionisation**?

....

....

1

(ii) State one instance where ionisation is:

A. **dangerous** to **human health**:

....

B. **useful** for **medical treatment**:

....

2

6. In the human eye, focusing of light rays on the retina is carried out by the cornea and lens.

(a) (i) Both lenses in a pupil's sunglasses have a power of +4·0 D. Calculate the **focal length** of the lenses.

Space for working and answer

2

(ii) Is the pupil **short-sighted** or **long-sighted**? ....

1

(b) The diagram below shows rays of light entering a teacher's left eye.

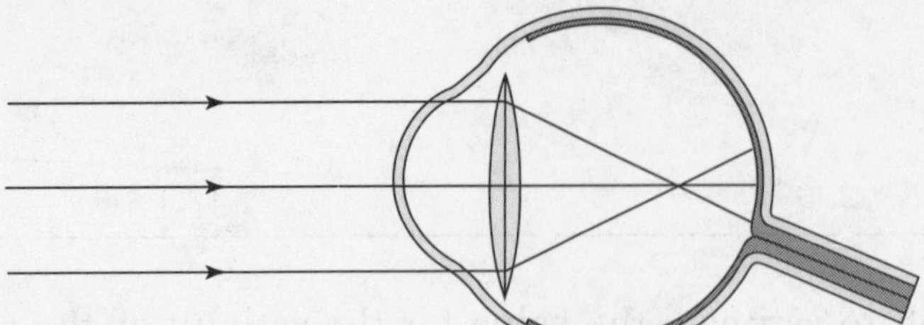

(i) Is the teacher **short-sighted** or **long-sighted**? .... | 1

(ii) Complete the diagram below to show how a **lens** can correct the teacher's eyesight.

You must draw the **correct type of lens in front of the eye** and show the **path of the three light rays on their journey through the lens to the back of the eye.**

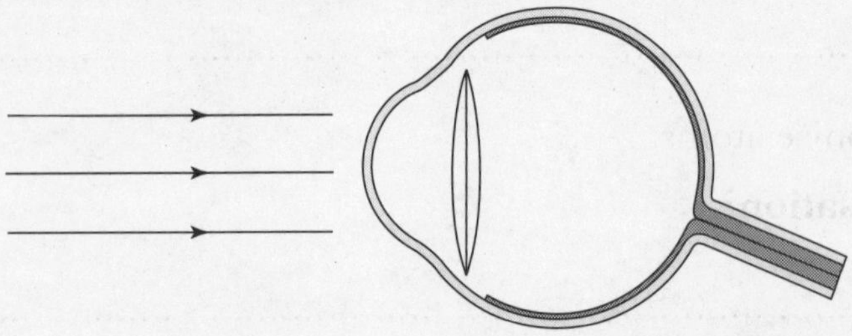

| | 2

(c) Recent television adverts have described how short sight can be corrected by laser surgery.

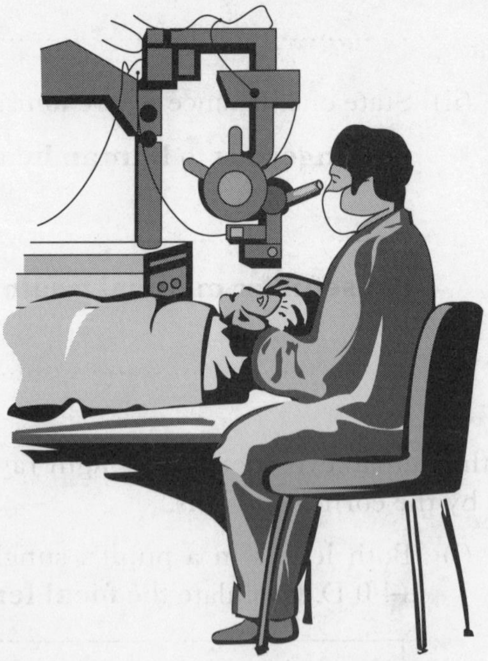

A laser is guided by a computer over the front surface of the eye (cornea). The laser fires pulses of light which last for a time of 2 ms. The light pulses burn away parts of the cornea, reshaping it. The reshaped cornea allows light rays to focus correctly on the back of the eye.

(i) State the effect laser surgery has on the focal length of the cornea.

....

.... | | 1

(ii) Suggest why pulses of laser light which last for a time **greater** than 2 ms are not used for eye surgery.

....

.... | | 1

	KU	PS

7. Physicists assign an **annual dose equivalent** value to every type of radiation.

The **annual dose equivalent** value is a measure of the damage radiation can do to the cells in a person's body if that person is exposed continuously to that radiation for one year.

The table shows the **annual dose equivalent** values at different places on earth due to cosmic radiation which reaches earth from outer space.

Place on earth	*Height above sea level (m)*	*Annual dose equivalent for cosmic radiation (mSv)*
sea level	0	0·25
summit of Ben Nevis mountain	1 343	0·34
summit of Mount Everest mountain	8 800	1·30
typical flight path of passenger jet aircraft	9 000	1·35

(a) Annual dose equivalent increases by 20% for every km increase in height above sea level.

Calculate the annual dose equivalent at a height of **10 km** above sea level.

> *Space for working and answer*

2

(b) Suggest why pilots of passenger aircraft are not allowed to fly for more than a specified period of time each year.

....

....

1

8. The electronic circuit for a shop security alarm is shown.

		KU	PS

- When it is dark, the logic level at U is 0. When it becomes light, the logic level changes to 1.
- When a window is opened, the logic level at V changes from 0 to 1.
- When the master switch is closed, the logic level at Y changes from 0 to 1.

(a) (i) Identify logic gate P: **1**

    (ii) Identify logic gate Q: **1**

    (iii) Complete this truth table for the shop security alarm circuit:

U	V	W	X	Y	Z
0	0			0	
0	1			0	
1	0			0	
1	1			0	
0	0			1	
0	1			1	
1	0			1	
1	1			1	

3

(b) (i) The security alarm designer decides that the master switch is **not** required.

    If the master switch is removed from the alarm circuit, state **one** other adjustment that must be made to the circuit in order for the circuit to function.

    ....

    .... **1**

    (ii) Suggest a suitable **input device** for the light sensor circuit.

    .... **1**

9. A pupil plans to build the following circuit to use as a temperature sensor in her fish tank.

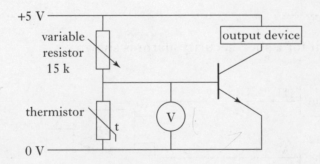

(a) The pupil wishes to use an output device which produces an **audible** warning when the temperature of the water in the fish tank becomes too low. Suggest a suitable **output device**.

    .... **1**

(b) (i) The transistor will start to conduct when the reading on the voltmeter rises to 0·7 V.

Calculate the voltage across the variable resistor when the transistor starts to conduct.

> *Space for working and answer*

1

(ii) Calculate the resistance of the thermistor when the transistor starts to conduct.

> *Space for working and answer*

2

(c) State the function of the variable resistor in the circuit.

....

....

1

(d) Explain how the pupil's circuit will operate to indicate when the temperature of the water in her fish tank becomes too low.

....

....

....

....

....

2

10. A *Keirin sprint* is a cycle race which takes place on an oval cycle track.

Cyclists are required to follow closely behind a motorcycle which originally travels at constant speed, before gradually accelerating.

The motorcycle then leaves the track and the cyclists have to sprint to the finish line, increasing their acceleration in the process.

(a) The graph shows the motion of a cyclist taking part in a *Keirin sprint* from the moment the motorcycle he is following starts to accelerate to the moment the cyclist crosses the finish line.

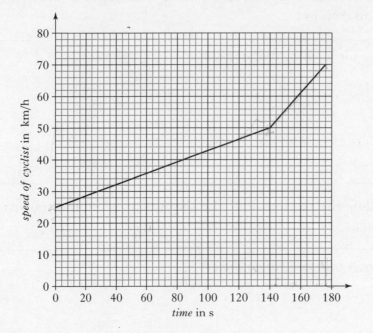

(i) What was the **increase** in speed of the cyclist from the moment the motorcycle he was following started to accelerate to the moment the motorcycle left the track?

Space for working and answer

(ii) Calculate the acceleration of the cyclist in **km/h/s** from the moment the motorcycle he was following left the track to the moment the cyclist crossed the finish line.

Space for working and answer

1

2

(b) The diagram shows the horizontal forces acting on the cyclist at a particular moment during the race.

frictional forces = 255·4 N

force exerted by cyclist on pedals = 266·6 N

The combined mass of the cyclist and his cycle is 68·0 kg.

Calculate the acceleration of the cyclist in **m/s²** at the particular moment.

Space for working and answer

3

11. A goalkeeper catches a football.

(a) The mass of the football is 0·45 kg. When the football reaches the goalkeeper's hands, it has a kinetic energy of 175 J.

Calculate the speed of the football at the instant it reaches the goalkeeper's hands.

Space for working and answer

2

(b) (i) The goalkeeper's hands bring the football to rest.

How much work must the hands do on the football to bring it to rest?

....

1

(ii) When the goalkeeper's hands bring the football to rest, the force of the impact causes the hands to move backwards a distance of 12·5 cm.

Calculate the size of the **force** the football must exert on the goalkeeper's hands as they bring the football to rest.

KU	PS

Space for working and answer

2

12. (*a*) Hydroelectricity is a renewable energy source.

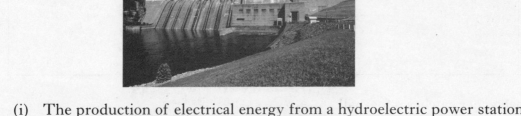

(i) The production of electrical energy from a hydroelectric power station can be started and stopped very quickly.

Explain why this is an **advantage**.

....

....

1

(ii) When hydro electric power stations are built, large areas of the surrounding land have to be permanently flooded with water.

Explain why this is a **disadvantage**.

....

....

1

(*b*) Coal and uranium are non-renewable sources of energy.

coal

uranium

One advantage a nuclear power station has over a coal-fired power station is that 1 kg of uranium produces the same amount of energy as 150 000 kg of coal.

If 1 kg of coal produces 28 MJ of energy, calculate how much energy 1 kg of uranium produces.

> *Space for working and answer*

1

13. A student sets up the apparatus shown below to determine the specific latent heat of fusion of water.

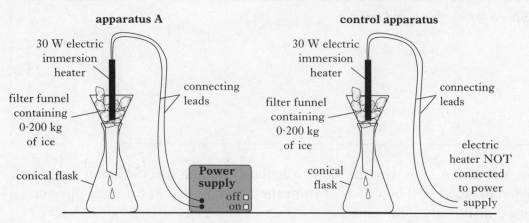

The electric immersion heater in apparatus A was switched on for a set time, then switched off. During this set time, some ice in both filter funnels melted and the liquid water was collected in the conical flasks.

The student took the following readings/measurements:

- *Power rating of electric immersion heater ..... 30.0 W*
- *Time interval between switching electric heater on and off ..... 240 s*
- *Mass of liquid water collected in conical flask of apparatus A while electric immersion heater was switched on ..... 0.218 kg*
- *Mass of liquid water collected in conical flask of control apparatus while electric immersion heater was switched on ..... 0·198 kg*

(a) Explain why the **control apparatus** is necessary.

....

....

....

2

Marks

KU	PS

(b) Calculate the **mass** of ice which is melted due to the electric immersion heater being switched on.

> *Space for working and answer*

1

(c) Use the student's readings/measurements to calculate a value for the specific latent heat of fusion of water.

> *Space for working and answer*

3

(d) Suggest **one** reason why the value calculated for the specific latent heat of fusion of water will be different from the value quoted in the data sheet on page 2 of this question paper.

....

....

1

14. The star Betelgeuse is 500 light-years from earth.

(a) Explain the term **light-year**.

....

....

1

(b) Betelgeuse emits X-rays, infrared radiation and visible light. These are all part of the electromagnetic spectrum.

(i) List these radiations in order of **increasing** wavelength.

....

1

(ii) State a suitable detector for:

X-rays: ....

Infrared radiation: ....

2

	KU	PS

(c) Visible light from Betelgeuse can be collected by and observed through an optical telescope.

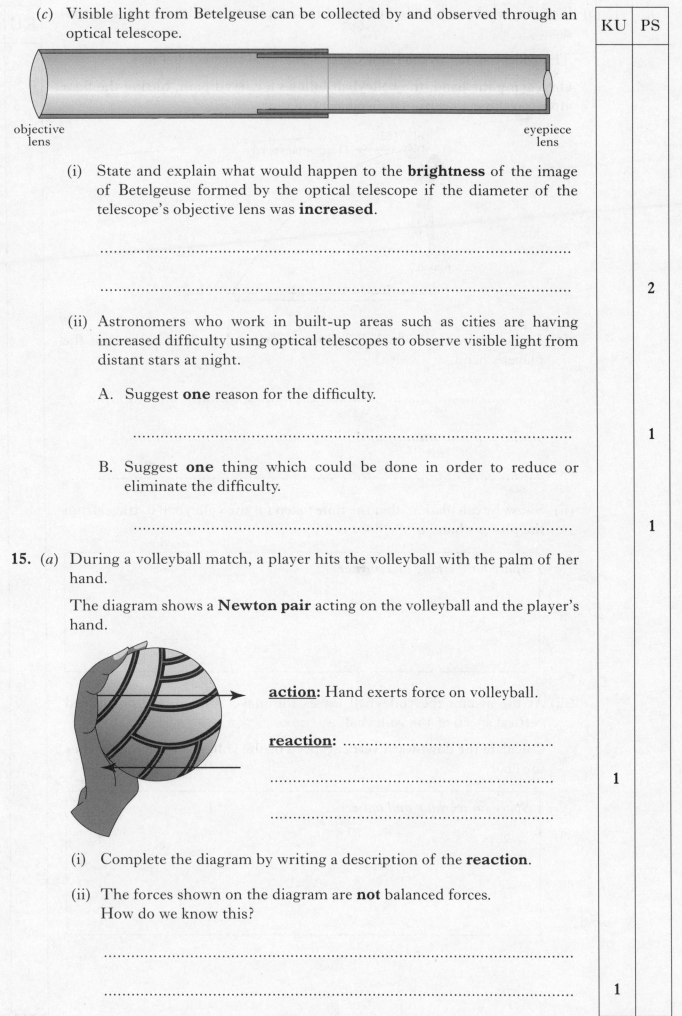

objective lens

eyepiece lens

(i) State and explain what would happen to the **brightness** of the image of Betelgeuse formed by the optical telescope if the diameter of the telescope's objective lens was **increased**.

....

....

2

(ii) Astronomers who work in built-up areas such as cities are having increased difficulty using optical telescopes to observe visible light from distant stars at night.

A. Suggest **one** reason for the difficulty.

....

1

B. Suggest **one** thing which could be done in order to reduce or eliminate the difficulty.

....

1

15. (a) During a volleyball match, a player hits the volleyball with the palm of her hand.

The diagram shows a **Newton pair** acting on the volleyball and the player's hand.

action: Hand exerts force on volleyball.

reaction: ....

....

....

1

(i) Complete the diagram by writing a description of the **reaction**.

(ii) The forces shown on the diagram are **not** balanced forces.
How do we know this?

....

....

1

(b) The player's hand hits the volleyball from a point directly above the central net.

The volleyball leaves the hand with a horizontal speed of 8·4 m/s.

On leaving the hand, the volleyball follows a curved path, hitting the floor after travelling a horizontal distance of 6·3 m.

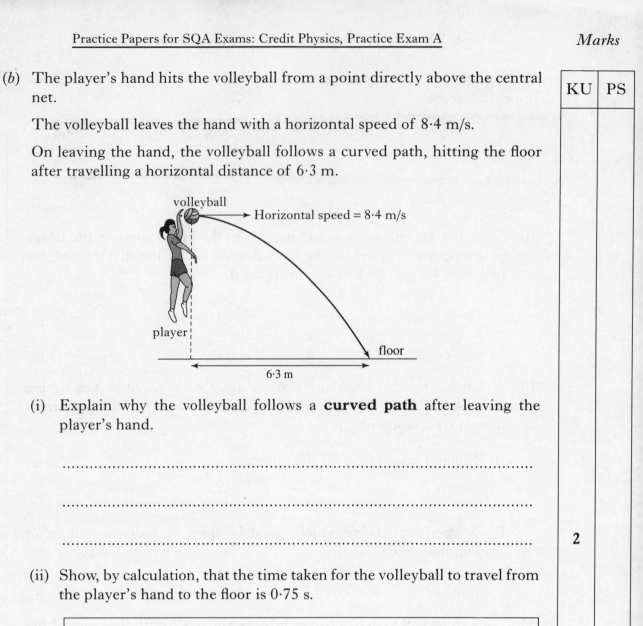

(i) Explain why the volleyball follows a **curved path** after leaving the player's hand.

....

....

....

2

(ii) Show, by calculation, that the time taken for the volleyball to travel from the player's hand to the floor is 0·75 s.

Space for working and answer

2

(iii) At the instant the volleyball leaves the player's hand, the downward vertical speed of the volleyball is 0 m/s.

Calculate the downward vertical speed of the volleyball when it reaches the floor.

Space for working and answer

2

Credit Physics

Practice Papers for SQA Exams	Time allowed: 1 hours, 45 minutes	Exam B

Fill in these boxes and read what is printed below.

Full name of school

Town

Forename(s)

Surname

Read each question carefully.

Attempt **all** questions. You may answer them in any order.

Write your answers in the spaces provided.

Write as neatly as possible. If you decide to change your answer, cross it out neatly before rewriting it.

Answer in sentences wherever possible.

Any data needed will be on the data sheet at the back of the book.

Use appropriate numbers of significant figures in final answer.

Leckie×Leckie
Scotland's leading educational publishers

Marks

	KU	PS

1. (*a*) The table shows information about three different radio bands.

Radio band	Wavelength range (m)	Use
extra low frequency (ELF)	above 100 000	communication with submerged submarines
low frequency (LF)	1000–10 000	radio broadcasts
ultra high frequency (UHF)	0·1–1	television broadcasts

    (i) What does the use for the ELF band suggest about the ability of extra low frequency waves to reflect off the surface of seas/oceans?

    ....

                                                                     1

    (ii) Explain why low frequency (LF) radio reception is better than ultra high frequency (UHF) television reception in hilly parts of the country.

    ....

    ....   **2**

(*b*) In a radio transmitter, a low frequency audio wave is combined with a much higher frequency radio carrier wave to produce a signal for transmission.

    (i) Complete the diagram below to show the transmitted signal:

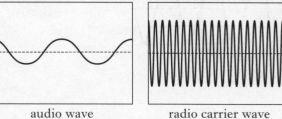

    audio wave   radio carrier wave   transmitted signal   **1**

    (ii) Name this process of combining the waves for transmission.

    ....   **1**

2. (*a*) A telephone conversation takes place between two people in St. Andrews and Edinburgh. The telephone lines in St.Andrews are made of **metal wire**. The telephone lines in Edinburgh are made of **optical fibre**.

St. Andrews

Edinburgh

(i) The table below lists some properties of **metal wires** and **optical fibres**. For each row in the right hand column of the table, **circle** the correct option – **metal wires** or **optical fibres**:

Property	Correct option
better signal quality	metal wires/optical fibres
highest signal capacity	metal wires/optical fibres
highest signal speed	metal wires/optical fibres
lowest cost	metal wires/optical fibres

(ii) A. Complete this diagram to show how **light** travels in an **optical fibre**.

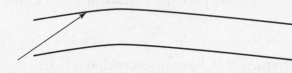

B. What name is given to this effect?

....

(iii) Traces **X** and **Y**, drawn to the same scale, show two different **musical notes** which are being transmitted along a **metal wire**.

Trace X   Trace Y

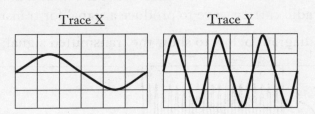

Explain which trace:

A. is **loudest**. ....

....

B. has the **highest frequency**. ....

....

(b) The telephone signals are transmitted back and forth between St. Andrews and Edinburgh in the form of microwaves using aerials fitted with **curved reflector dishes**.

curved reflector

aerial

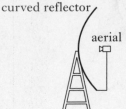

St. Andrews

curved reflector

aerial

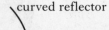

Edinburgh

KU: 2, 2, 1

PS: 2

Marks

KU	PS

(i) State **one** advantage of fitting curved reflector dishes to transmitter and receiver aerials.

....

1

(ii) A. At what speed do the microwaves travel through air?

....

1

B. The microwaves have a frequency of 1500 MHz. Calculate their **wavelength**.

Space for working and answer

2

3. A sound engineer sets up the apparatus shown to measure the speed of sound in air. The time taken for a pulse of sound to travel a measured distance between the two microphones is recorded on the electronic timer.

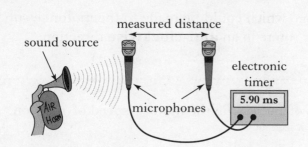

The sound engineer performs the experiment four times. Her readings are shown in this table:

Measured distance between microphones (m)	Time recorded on electronic timer (ms)
2·00	5·90
2·00	6·15
2·00	6·10
2·00	5·85

(a) Use **all** the times shown in the table to calculate a value for the speed of sound in air in **m/s**.

Space for working and answer

3

(b) Suggest **one** improvement the sound engineer could make to the **apparatus** in order to achieve a more accurate value for the speed of sound in air.

....

4. A simple electric motor is shown:

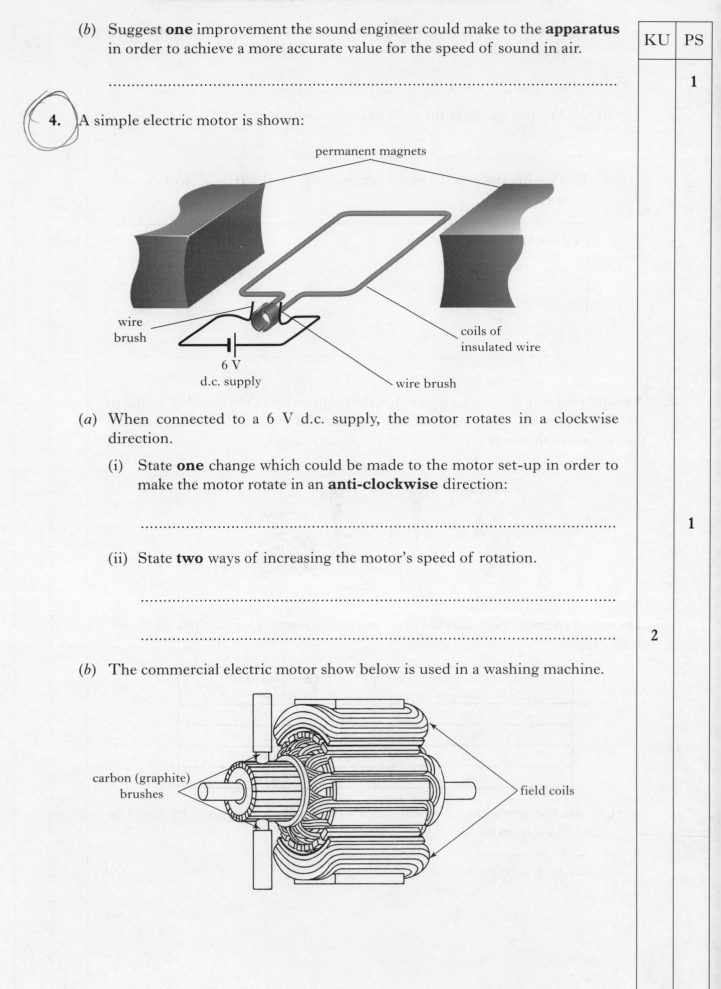

(a) When connected to a 6 V d.c. supply, the motor rotates in a clockwise direction.

(i) State **one** change which could be made to the motor set-up in order to make the motor rotate in an **anti-clockwise** direction:

....

(ii) State **two** ways of increasing the motor's speed of rotation.

....

....

(b) The commercial electric motor show below is used in a washing machine.

State **one** reason why each of the following are used in the commercial motor:

• **carbon (graphite) brushes:** ....

....

• **field coils:** ....

....

5. Pupils attend a lunchtime physics club where they carry out different electricity projects.

(*a*) For his project, one pupil requires a resistance of 220 Ω.

However, he only has the following three resistors available.

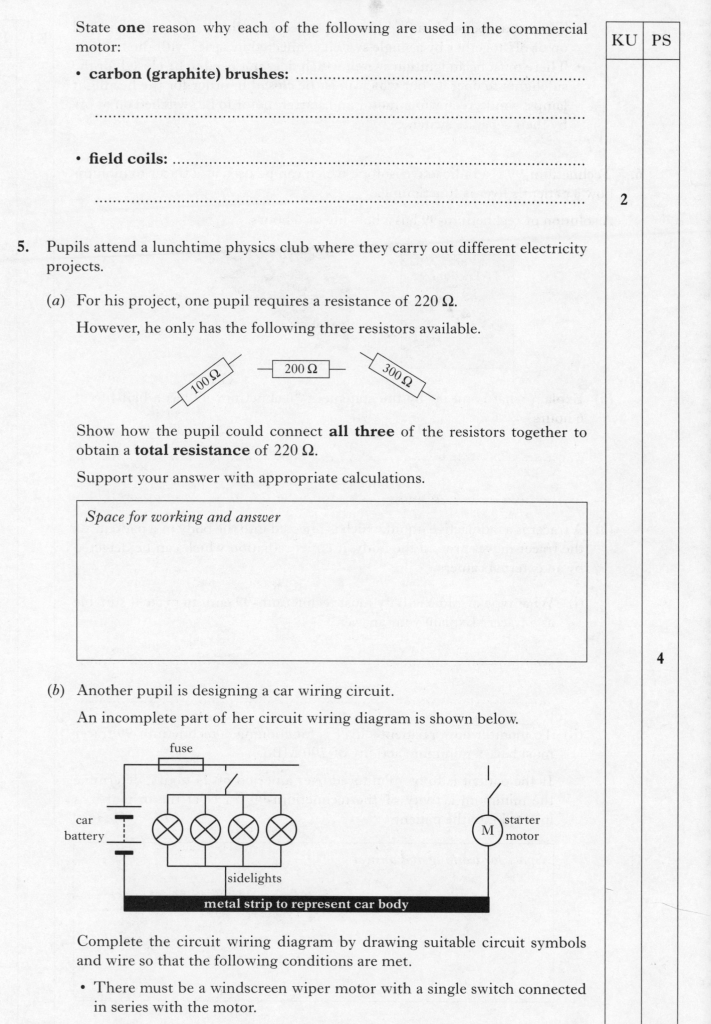

Show how the pupil could connect **all three** of the resistors together to obtain a **total resistance** of 220 Ω.

Support your answer with appropriate calculations.

> *Space for working and answer*

(*b*) Another pupil is designing a car wiring circuit.

An incomplete part of her circuit wiring diagram is shown below.

Complete the circuit wiring diagram by drawing suitable circuit symbols and wire so that the following conditions are met.

• There must be a windscreen wiper motor with a single switch connected in series with the motor.

KU | PS
2

4

- There must be two headlight lamps connected so that both can be switched on or off together by a single switch connected in series with the lamps.
- There must be an ignition switch which does not need to be closed for the sidelights to operate but which must be closed in order for the headlight lamps, windscreen wiper motor and starter motor to be switched on or off by their separate switches.

6. Technetium-99 is a radioactive isotope which can be used as a tracer to monitor how a patient's liver is functioning.

A solution of technetium-99 has a half-life of 6 hours.

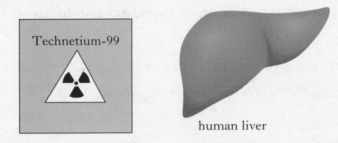

human liver

(a) Explain what is meant by the statement "technetium-99 has a half-life of 6 hours".

....

....

(b) A tracer is a radioactive liquid which is injected into the body of a patient. As the tracer moves around the body, it emits radiation which can be detected by an external camera.

(i) What type of radioactivity must technetium-99 emit to make it suitable as a tracer? Explain your answer.

....

....

(ii) To monitor how a patient's liver is functioning, a technetium-99 tracer must have a minimum activity of 200 MBq.

If the patient is to be monitored over a period of 24 hours, determine the minimum activity of the technetium-99 tracer at the instant it is injected into the patient.

> *Space for working and answer*

KU	PS
	3
1
	2
2

(c) When medical technicians prepare radioactive tracers, they must follow strict safety guidelines.

State **two** safety precautions the medical technicians should take while preparing a sample of technetium-99 tracer.

....

....

7. (a) A boy stands behind and to the left of a girl in a brightly-lit room. Both look into a plane mirror in front of them.

The diagram shows a ray of light travelling from the boy's face to the mirror, then entering the girl's eye. Many light rays follow a similar path, allowing the girl to see the boy's face.

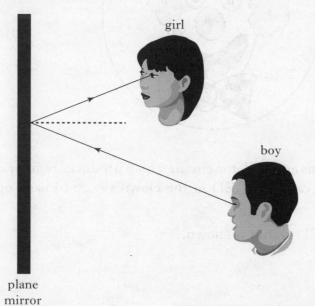

girl

boy

plane
mirror

Explain, using the **principle of reversibility of ray paths**, why the **boy** can also see a reflection of the girl's face in the mirror.

....

....

....

KU	PS
	6
	2
1	

(b) The diagram shows a ray of light travelling through air and striking the flat surface of a transparent glass block.

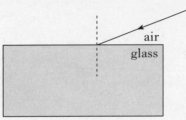

(i) On the diagram, draw the path taken by the ray of light as it enters the glass.

(Do not show the ray of light leaving the glass block.)

(ii) On the diagram, clearly show and label the angle of refraction.

(iii) What name is given to the dashed line which is drawn at 90° to the flat face of the glass block?

....

8. A novelty badge shows the face of a clown.

The badge contains an oscillator circuit which produces regular clock pulses. The clock pulses cause an LED in the clown's nose to flash on and off at regular intervals.

The oscillator/LED circuit is shown.

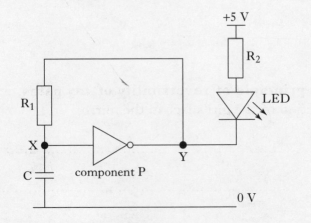

KU	PS
	1
1	
1	

Marks

	KU	PS

(a) Name component P.

.... | **1** | |

(b) By referring to points **X** and **Y** in the oscillator circuit, explain how the circuit produces a regular series of clock pulses.

You should assume that capacitor C is **discharged** at the start of the process.

....

....

....

.... | **2** | |

(c) Capacitor C is replaced by a capacitor with a higher capacitance.

State and explain the effect this will have on the frequency of the flashing LED.

....

....

.... | | **2** |

(d) When the LED is lit, the total voltage across the LED and protective resistor R_2 is $5 \cdot 0$ V.

The voltage across the LED is $0 \cdot 5$ V and the current through it is 10 mA. Calculate the resistance of the protective resistor R_2.

Space for working and answer

| | | **3** |

9. An electric guitar contains an amplifier.

The amplifier is connected to a loudspeaker.

The technical specification for the guitar-amplifier-loudspeaker system at a particular instant when the guitar is being played is shown below.

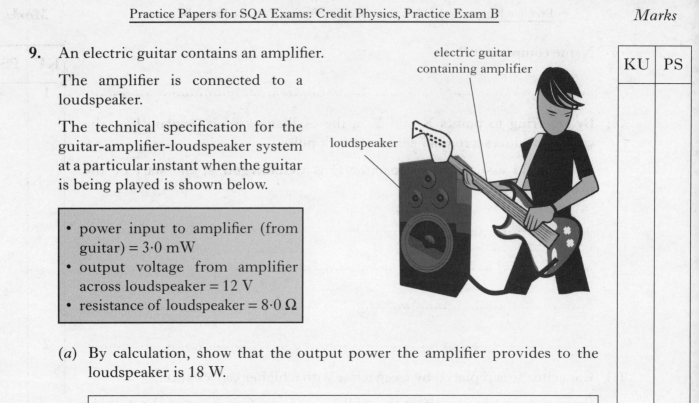

- power input to amplifier (from guitar) = 3·0 mW
- output voltage from amplifier across loudspeaker = 12 V
- resistance of loudspeaker = 8·0 Ω

(*a*) By calculation, show that the output power the amplifier provides to the loudspeaker is 18 W.

Space for working and answer

2

(*b*) Calculate the power gain of the amplifier.

Space for working and answer

2

(*c*) The guitarist wishes to check that the voltage gain control of the amplifier in his electric guitar is working correctly.

He removes the amplifier from his guitar and connects the amplifier to the apparatus shown.

KU	PS

Marks

KU	PS

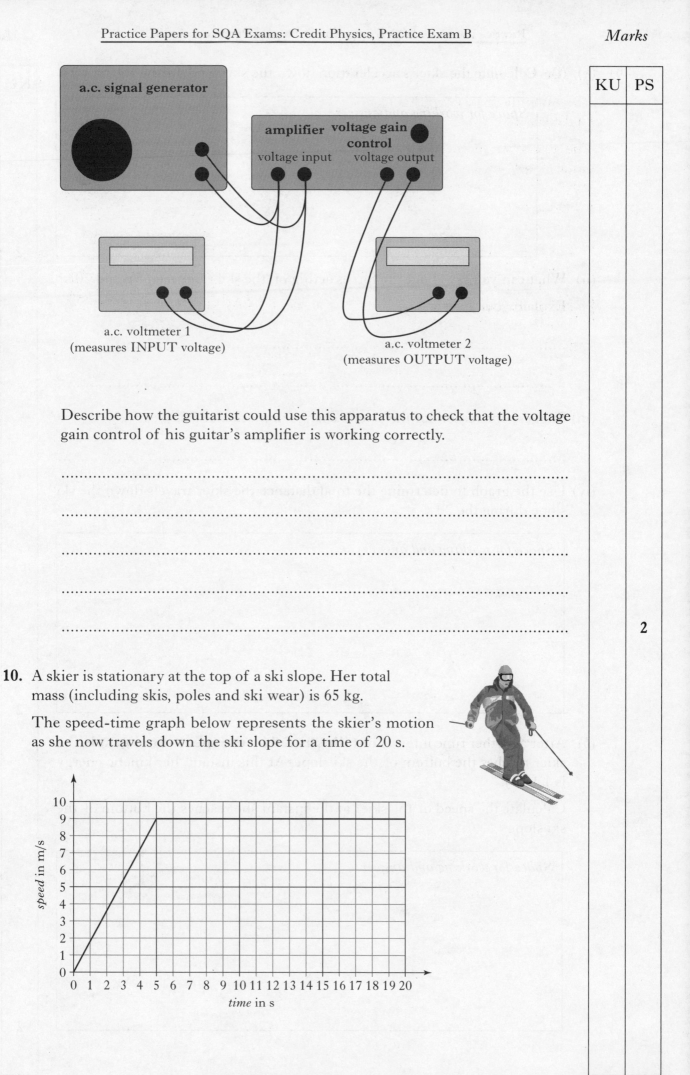

Describe how the guitarist could use this apparatus to check that the voltage gain control of his guitar's amplifier is working correctly.

....

....

....

....

....

2

10. A skier is stationary at the top of a ski slope. Her total mass (including skis, poles and ski wear) is 65 kg.

The speed-time graph below represents the skier's motion as she now travels down the ski slope for a time of 20 s.

Marks

KU	PS

(a) (i) Calculate the skier's acceleration down the ski slope during the first 5 s.

Space for working and answer

2

(ii) What can you say about the forces acting on the skier between 5 s and 20 s? Explain your answer.

....

....

2

(iii) Suggest one reason for the skier's change in motion at 5 s.

....

1

(iv) Use the graph to determine the total distance the skier travels down the ski slope during the 20 s.

Space for working and answer

2

(b) After a further time interval **not illustrated by the speed-time graph**, the skier reaches the bottom of the ski slope. At this instant, her kinetic energy is 1 170 J.

Calculate the speed of the skier at the instant she reaches the bottom of the ski slope.

Space for working and answer

2

11. A martial arts expert is performing at a karate demonstration. She breaks a block of wood in two by striking the wood with the edge of her hand.

The diagram below shows the forces acting on her arm and hand as they accelerate downwards towards the wood block.

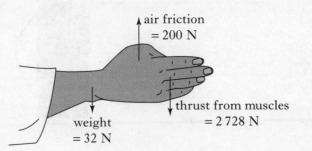

air friction = 200 N

thrust from muscles = 2 728 N

weight = 32 N

(a) Calculate the **size** of the unbalanced force acting on the arm and hand at the instant shown.

> *Space for working and answer*

2

(b) The arm and hand have a combined mass of 3·2 kg.

Calculate the **size** of their acceleration at the instant shown.

> *Space for working and answer*

2

(c) One part of the broken block of wood has a mass of 0·28 kg. It falls to the floor, through a height of 0·75 m.

Calculate the work done by gravity on this broken part of the block as it falls to the floor.

> *Space for working and answer*

2

KU | PS

12. A battery in a walkie-talkie is recharged by connecting the walkie-talkie to a transformer. The primary coil of the transformer is connected to a 230 V a.c. mains supply.

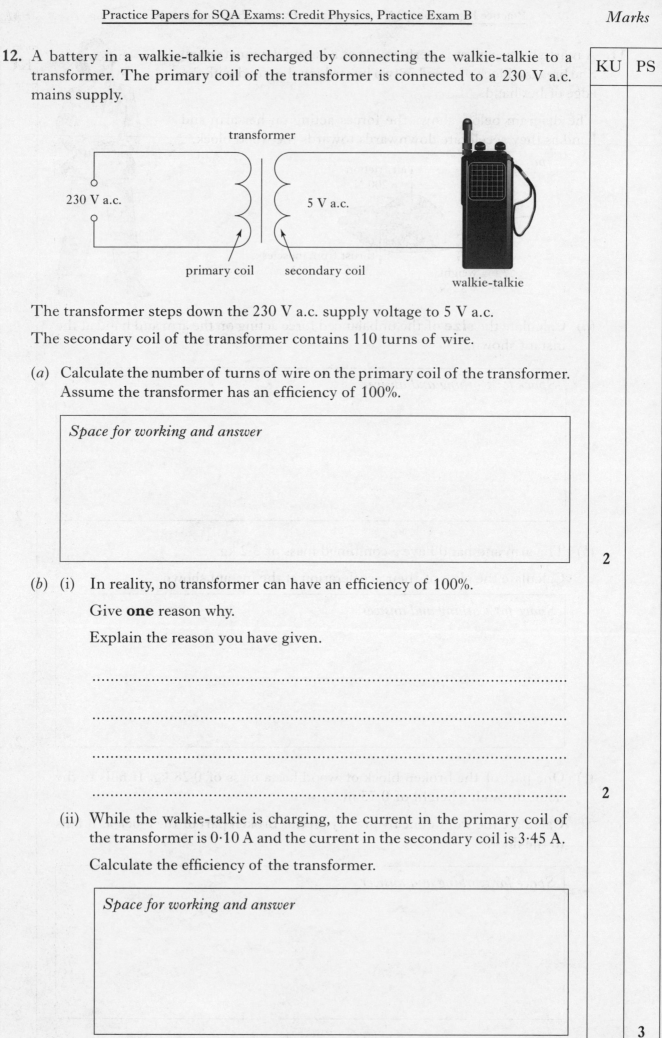

The transformer steps down the 230 V a.c. supply voltage to 5 V a.c.

The secondary coil of the transformer contains 110 turns of wire.

(*a*) Calculate the number of turns of wire on the primary coil of the transformer. Assume the transformer has an efficiency of 100%.

> *Space for working and answer*

2

(*b*) (i) In reality, no transformer can have an efficiency of 100%.

Give **one** reason why.

Explain the reason you have given.

....

....

....

....

2

(ii) While the walkie-talkie is charging, the current in the primary coil of the transformer is 0·10 A and the current in the secondary coil is 3·45 A.

Calculate the efficiency of the transformer.

> *Space for working and answer*

3

KU | PS

(c) Inside the walkie-talkie, there is a component which changes the current supplied by the transformer from a.c. to d.c. This allows the charging current to flow through the battery in a single direction.

Draw the circuit symbol for this component.

Space for answer

1

13. A student is provided with an electronic thermometer and a small glass beaker containing 150 g of hot, liquid wax.

To investigate the cooling of the wax, the student sets up the apparatus as shown.

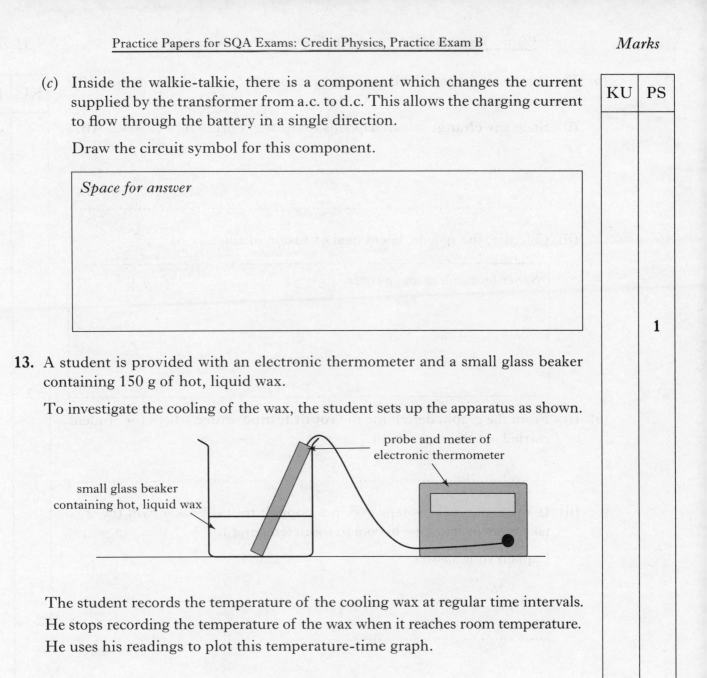

The student records the temperature of the cooling wax at regular time intervals. He stops recording the temperature of the wax when it reaches room temperature. He uses his readings to plot this temperature-time graph.

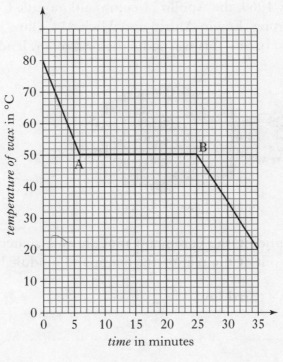

Marks

(a) During the time interval AB, the wax gives out 40 500 J of heat energy to the surroundings.

(i) State any **change** which happens to the wax during time interval AB.

....

....

(ii) Calculate the specific latent heat of fusion of the wax.

Space for working and answer

(b) (i) From the graph, determine the **room temperature** where the student carried out the experiment.

....

(ii) If the experiment is repeated in a **copper metal** beaker, will the wax take more or less time to cool to room temperature?

Explain your answer.

....

....

14. Between 16 and 24 July 1969, the Apollo 11 command module Columbia carried astronauts Neil Armstrong, Edwin Aldrin and Michael Collins on their historic voyage to the moon and back – this was to be the first moon landing.

command module Colombia

(a) Whenever the engine of the command module was switched on, exhaust gases were pushed out of the rear of the command module by the engine.

(i) State **Newton's third law**.

....

....

KU | PS

1

2

1

2

1

(ii) Use **Newton's third law** to explain how the command module was propelled **forward** due to the emission of the exhaust gases.

....

....

(b) The astronauts orbited the moon in the command module. The engines of the command module were switched off. The orbit was a circular path.

Explain why the command module followed a circular path around the moon.

....

....

....

....

(c) Astronauts Armstrong and Aldrin descended to the moon's surface in a lunar module called Eagle.

Lunar module Eagle

On earth, the weight of the lunar module (including both astronauts and the same fuel load) was 104 760 N.

Determine the mass of the lunar module on the **moon's** surface. Explain your answer.

Space for working and answer

KU	PS
1	
	2
	3

Marks

KU	PS

15. Optical telescopes are used to observe stars from the earth.

A diagram of an optical telescope is shown.

objective
lens

eyepiece
lens

(a) (i) State the purpose of the objective lens in this telescope.

....

....

1

(ii) Explain why the objective lens should have as **large** a diameter as possible.

....

....

1

(iii) The focal length of the objective lens in this telescope is 500 mm. Calculate the power of the objective lens.

Space for working and answer

2

(b) The eyepiece lens of the optical telescope is used as a magnifying lens.

Complete the diagram to show how the lens forms a magnified image of the object.

Your diagram must clearly show the position and size of the image formed.

focus object   focus

magnifying lens

3

Practice Exam C

Credit Physics

Practice Papers for SQA Exams	Time allowed: 1 hours, 45 minutes	**Exam C**

Fill in these boxes and read what is printed below.

Full name of school

Town

Forename(s)

Surname

Read each question carefully.

Attempt **all** questions. You may answer them in any order.

Write your answers in the spaces provided.

Write as neatly as possible. If you decide to change your answer, cross it out neatly before rewriting it.

Answer in sentences wherever possible.

Any data needed will be on the data sheet at the back of the book.

Use appropriate numbers of significant figures in final answer.

Leckie ✕ Leckie
Scotland's leading educational publishers

1. (*a*) The diagram illustrates television transmission.

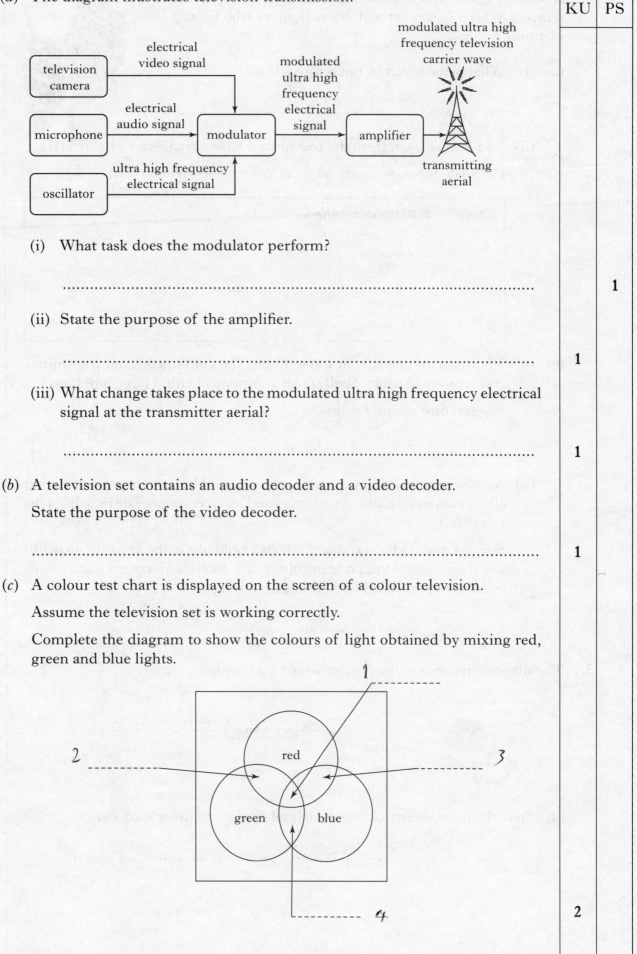

| | | KU | PS |

(i) What task does the modulator perform?

.... **1**

(ii) State the purpose of the amplifier.

.... **1**

(iii) What change takes place to the modulated ultra high frequency electrical signal at the transmitter aerial?

.... **1**

(*b*) A television set contains an audio decoder and a video decoder.

State the purpose of the video decoder.

.... **1**

(*c*) A colour test chart is displayed on the screen of a colour television.

Assume the television set is working correctly.

Complete the diagram to show the colours of light obtained by mixing red, green and blue lights.

2

2. A security guard uses a combined radio transmitter and receiver to keep in contact with his colleagues who have identical devices.

(a) (i) What is the speed of radio waves in air?

....

1

(ii) Radio waves sent from the transmitter have a frequency of 400 MHz. Calculate the wavelength of these radio waves in air.

> *Space for working and answer*

2

(b) (i) The security guard is **not** able to use the combined radio transmitter and receiver to contact colleagues positioned behind large buildings.

Suggest **one** reason for this.

....

1

(ii) A radio engineer suggests this problem could be overcome by adjusting all the combined radio transmitters and receivers to operate on a different waveband.

Suggest **one** radio waveband which would allow the security guard to use the combined radio transmitter and receiver to communicate with colleagues positioned behind large buildings.

....

1

3. The diagram represents the transmission of a telephone signal.

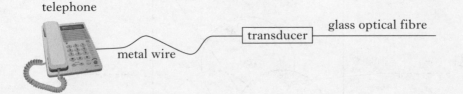

(a) State the main energy change which takes place in the transducer.

....

1

(*b*) The telephone signal travels most of its journey through the optical fibre, not the metal wire.

State **two** advantages optical fibres have over metal wires for the transmission of telephone signals.

....

....

(*c*) This diagram shows two light rays, A and B, entering and leaving a short section of glass optical fibre.

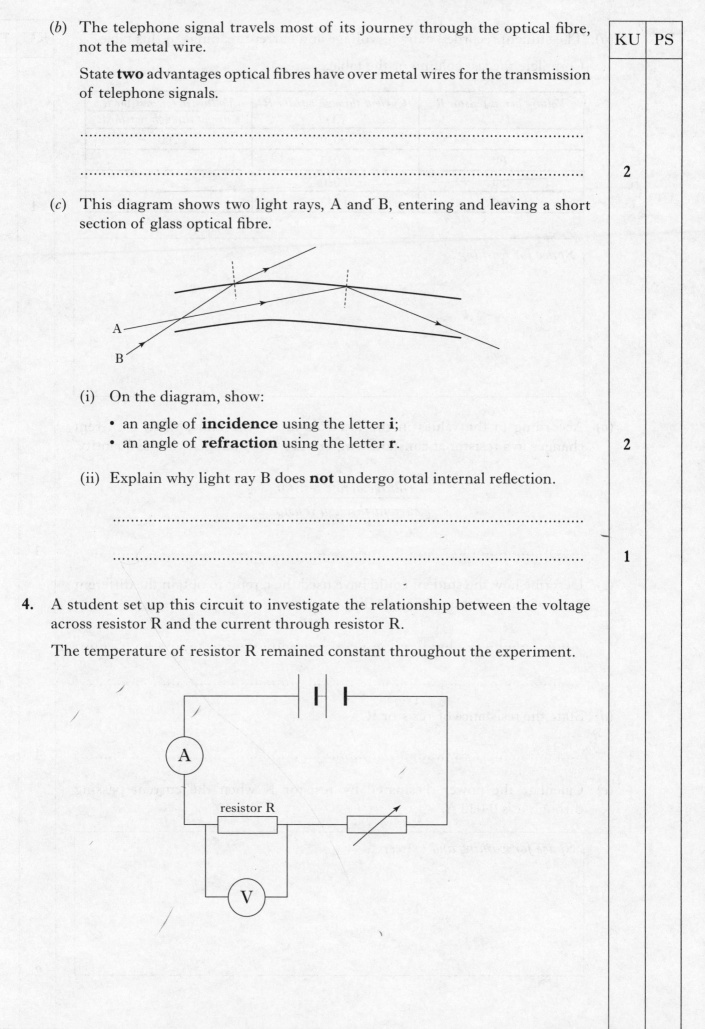

   (i)   On the diagram, show:

       • an angle of **incidence** using the letter **i**;
       • an angle of **refraction** using the letter **r**.

   (ii) Explain why light ray B does **not** undergo total internal reflection.

....

....

4. A student set up this circuit to investigate the relationship between the voltage across resistor R and the current through resistor R.

The temperature of resistor R remained constant throughout the experiment.

resistor R

KU | PS

2

2

1

		KU	PS

(a) The student recorded pairs of voltage and current readings in this table.

Complete the last column of the table.

Voltage across resistor R (V)	Current through resistor R (A)	$\dfrac{\text{Voltage across resistor } R}{\text{Current through resistor } R}$
6·0	0·030	200
7·0	0·035	
8·0	0·040	
9·0	0·045	

Space for working

1

(b) According to the values in the last column of the table, when a current changes in a resistor at constant temperature, what happens to the quantity

$$\frac{\text{voltage across resistor}}{\text{current through resistor}} \, ?$$

....

1

(c) Describe how the student could have used the circuit to obtain the different pairs of current and voltage readings.

....

....

1

(d) State the resistance of resistor R.

....

1

(e) Calculate the power dissipated by resistor R when the current passing through it is 0·030 A.

Space for working and answer

2

Marks

KU	PS

5. One kilowatt-hour (kWh) is the quantity of electrical energy transferred by an electrical appliance with a power rating of 1 kW when it is switched on for 1 hour.

(*a*) Show, by calculation, that **1 kWh = 3 600 000 J.**

> *Space for working and answer*

2

(*b*) Four electrical appliances, with the power ratings shown, are connected to the mains supply of a home.

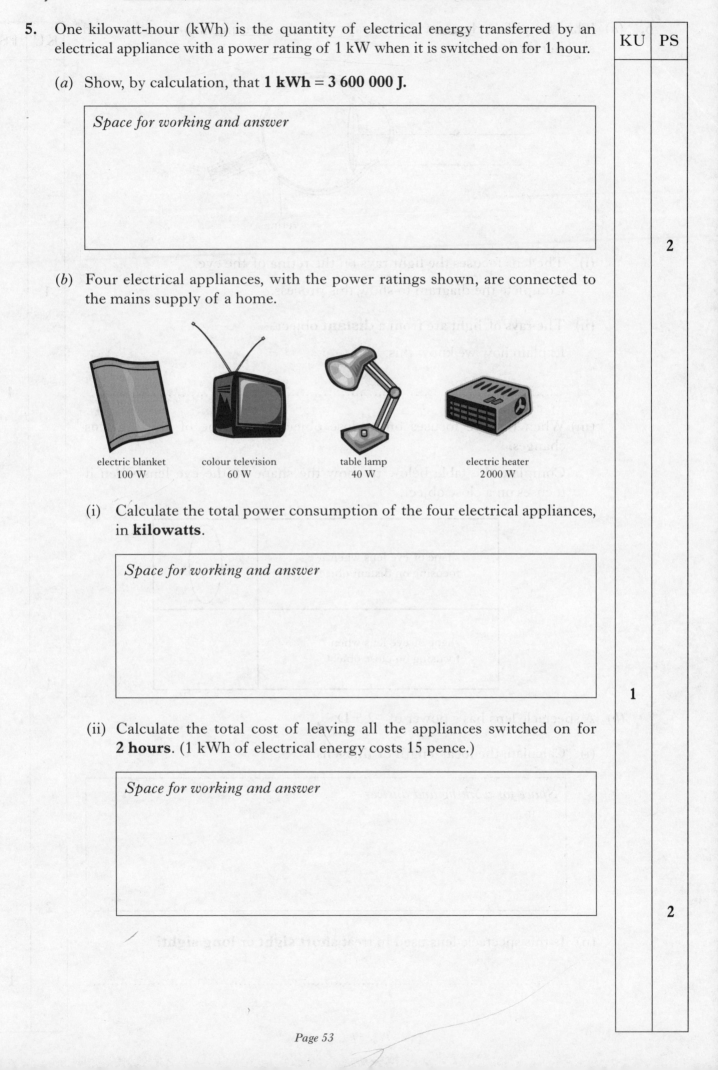

electric blanket 100 W   colour television 60 W   table lamp 40 W   electric heater 2 000 W

(i) Calculate the total power consumption of the four electrical appliances, in **kilowatts**.

> *Space for working and answer*

1

(ii) Calculate the total cost of leaving all the appliances switched on for **2 hours**. (1 kWh of electrical energy costs 15 pence.)

> *Space for working and answer*

2

6. (*a*) This simplified diagram shows rays of light entering a human eye and striking the lens.

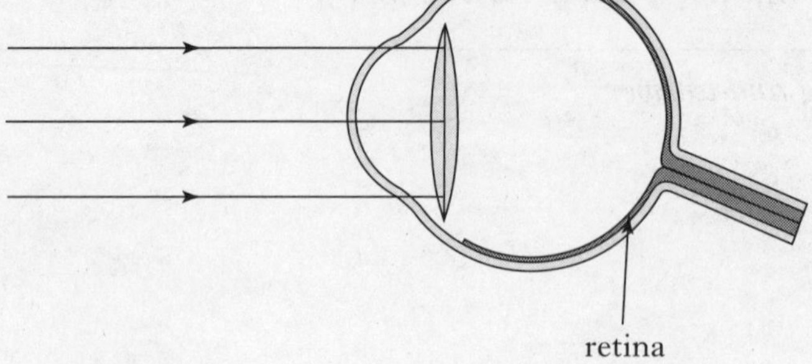

retina

(i) The lens focuses the light rays on the retina of the eye.

Complete the diagram to show this process.

1

(ii) The rays of light are from a **distant** object.

Explain how we know this.

....

1

(iii) When the eye focuses on a close object, the shape of the eye lens changes.

Complete the table below to show the shape of the eye lens when it focuses on a close object.

shape of eye lens when focusing on distant object	
shape of eye lens when focusing on close object	

1

(*b*) A spectacle lens has a power of −2·5 D.

(i) Calculate the focal length of this lens.

Space for working and answer

2

(ii) Is this spectacle lens used to treat **short sight** or **long sight**?

....

1

Marks

	KU	PS

7. Ultrasound and X-rays can be used to examine inside the human body.

   (*a*) (i) State what is meant by the term ultrasound.

....

....

      (ii) Describe how pulses of ultrasound can be used to build up an image of kidney stones in a patient's kidney.

....

....

....

....

     (iii) What is the speed of ultrasound in human body tissue?

....

     (iv) Calculate the time taken for a pulse of ultrasound to travel through 3 cm of tissue.

> *Space for working and answer*

   (*b*) (i) State one suitable detector for X-rays.

....

      (ii) Suggest why X-rays are not used to examine body organs, such as kidneys, for prolonged periods of time.

....

....

Marks column values: KU 1 (a)(i), PS 2 (a)(ii), KU 1 (a)(iii), PS 2 (a)(iv), KU 1 (b)(i), 1 (b)(ii)

(iii) X-rays are used to treat cancer tumours.

A beam of X-rays is fired at a tumour inside a patient's body from a machine which continually rotates around the body.

Explain why the tumour receives a large dose of X-rays but the surrounding tissues receive a much smaller dose.

....

....

....

2

8. (a) Two logic gate truth tables are shown.

Identify the type of logic gate represented by each truth table.

(i)

input A	input B	output
0	0	0
0	1	1
1	0	1
1	1	1

Type of logic gate:

....

(ii)

input A	input B	output
0	0	0
0	1	0
1	0	0
1	1	1

Type of logic gate:

....

2

(b) A simple, electronically-controlled padlock has three push-button switches, A, B and C. In order to open the padlock, the correct combination of switches must be closed.

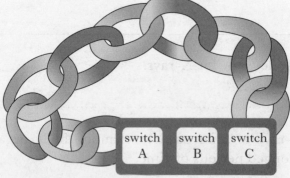

The logic circuit for the padlock is shown.

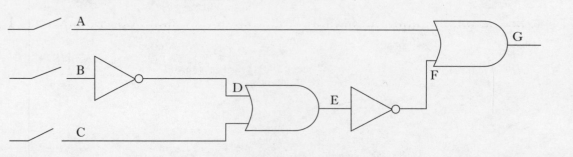

		KU	PS

When a switch is **open**, its logic output is **0**.

When a switch is **closed**, its logic output is **1**.

The padlock will only open when the **output G** has a logic level of **1**.

(i) Complete this truth table for the logic circuit to show which switch or switches must be closed in order to open the padlock.

A	B	C	D	E	F	G
0	0	0	1			
0	0	1	1			
0	1	0	0			
0	1	1	0			
1	0	0	1			
1	0	1	1			
1	1	0	0			
1	1	1	0			

3

(ii) Explain why a **solenoid** is a suitable **output device** for this logic circuit.

....

....

1

9. A teacher uses a novelty electronic egg timer to tell him when his breakfast egg has boiled for a set time.

The egg timer has a timing circuit which displays time in seconds.

When the time displayed on the egg timer reaches the value set by the teacher, the egg timer plays a tune via an internal amplifier-loudspeaker circuit.

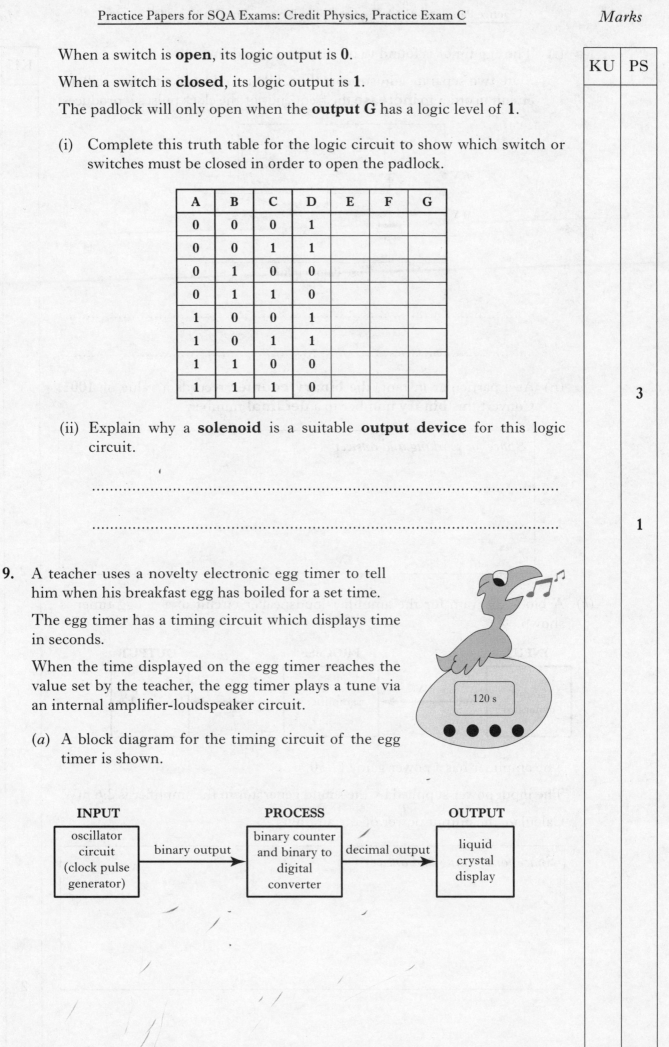

120 s

(a) A block diagram for the timing circuit of the egg timer is shown.

INPUT		PROCESS		OUTPUT
oscillator circuit (clock pulse generator)	binary output →	binary counter and binary to digital converter	decimal output →	liquid crystal display

(i) The egg timer is found to be running behind time.

State **two** separate adjustments the teacher could make to the oscillator circuit in order to **increase** the frequency of the clock pulses it produces.

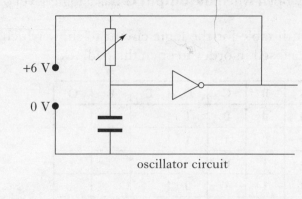

oscillator circuit

....

....

2

(ii) At a particular instant, the **binary counter** records a value of **1001**. Convert this **binary** number to a **decimal** number.

Space for working and answer

2

(b) A block diagram for the amplifier-loudspeaker circuit of the egg timer is shown.

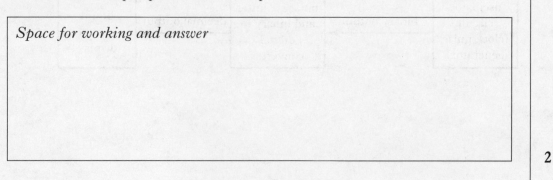

The amplifier has a power gain of 750.

The input power supplied by the sound generator to the amplifier is 3·6 mW.

Calculate the output power of the amplifier.

Space for working and answer

2

10. A skydiver jumps from an aircraft travelling at a height of 4 000 m.

After free-falling through the air, she opens her parachute when she is 1 200 m above the ground.

The combined mass of the skydiver and her parachute is 72·0 kg.

(a) (i) Calculate the **decrease** in gravitational potential energy of the skydiver during this part of her journey.

Space for working and answer

(ii) At the instant the skydiver leaves the aircraft, her downward speed is 0 m/s.

Calculate the speed of the skydiver at the instant before she opens her parachute.

Space for working and answer

(b) The graph shows how the speed of the skydiver changes from the instant she leaves the aircraft to the instant she lands on the ground.

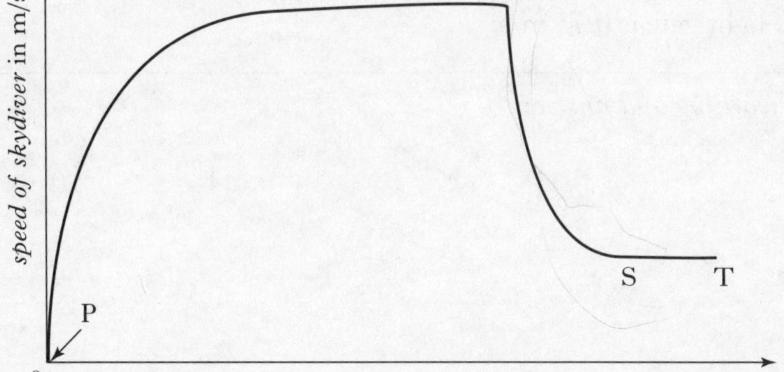

3

3

(i) Which point on the graph (**P**, **Q**, **R**, **S** or **T**) represents the instant the skydiver opens her parachute?

....

(ii) During which **two** sections of the graph are the forces acting on the skydiver **unbalanced**?

....

(c) During free fall, before opening her parachute, suggest **one** way the skydiver could **reduce** the **frictional force** acting on her.

....

11. A pupil aims to determine the acceleration of a toy car as it runs down a slope.

The toy car is released from rest at the top of the slope.

As the toy car accelerates down the slope, the mask passes through the light gate. The time taken for the mask to pass through the light gate is recorded on the electronic timer.

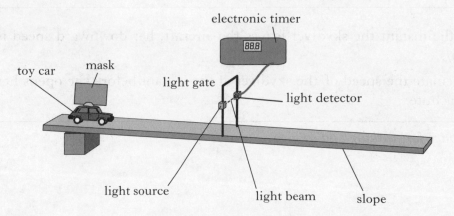

The readings obtained for one experiment are shown in the table.

Speed of toy car at top of slope (m/s)	Length of mask (cm)	Time taken for mask to pass through light gate (s)	Speed of toy car as mask passes through light gate (m/s)	Time taken for toy car to travel from top of slope to light gate (s)
0	6·0	0·08	0·75	0·50

(a) (i) Show, by calculation, that the speed of the toy car as the mask passes through the light gate is 0·75 m/s.

Space for working and answer

KU	PS
	1
	1
1	
2	

Marks

KU	PS

(ii) Using information from the table, calculate the acceleration of the toy car as it travels from the top of the slope to the light gate.

> *Space for working and answer*

2

(b) The pupil uses a stopwatch to measure the time taken for the toy car to travel from the top of the slope to the light gate.

Explain why this is likely to lead to an **inaccurate** time measurement.

....

....

1

(c) To obtain a value closer to the *true* **instantaneous** speed of the toy car as the mask passes through the light gate, should the **length** of the mask be **increased** or **decreased**?

Explain your answer.

....

....

....

2

Marks

	KU	PS

12. The output from a simple a.c. generator is connected to an oscilloscope.

The simple a.c. generator consists of a cylindrical bar magnet and a coil of copper wire which is wound around an iron core. The bar magnet is rotated next to the coil of wire.

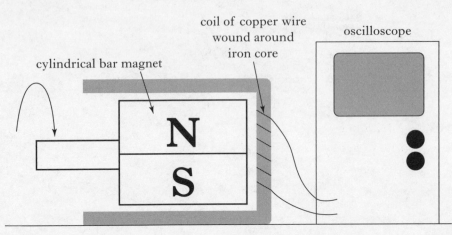

coil of copper wire
wound around
iron core

oscilloscope

cylindrical bar magnet

N

S

(*a*) (i) Explain why a voltage is induced across the coil of copper wire when the bar magnet is rotated.

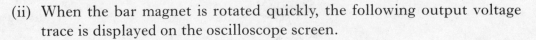

1

(ii) When the bar magnet is rotated quickly, the following output voltage trace is displayed on the oscilloscope screen.

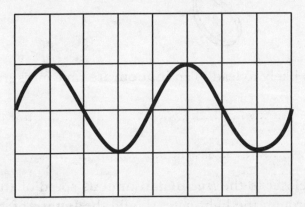

On the same diagram, draw a possible output voltage trace for when the bar magnet is rotated at a **lower** speed.

Assume no adjustment has been made to any controls on the oscilloscope.

2

(iii) The simple a.c. generator is adapted by winding more turns of copper wire around the iron core.

Assuming a constant speed of rotation for the bar magnet before and after the change, state how the size of the voltage induced across the coil of wire will be affected.

1

(b) A wind turbine contains a full size a.c. generator.

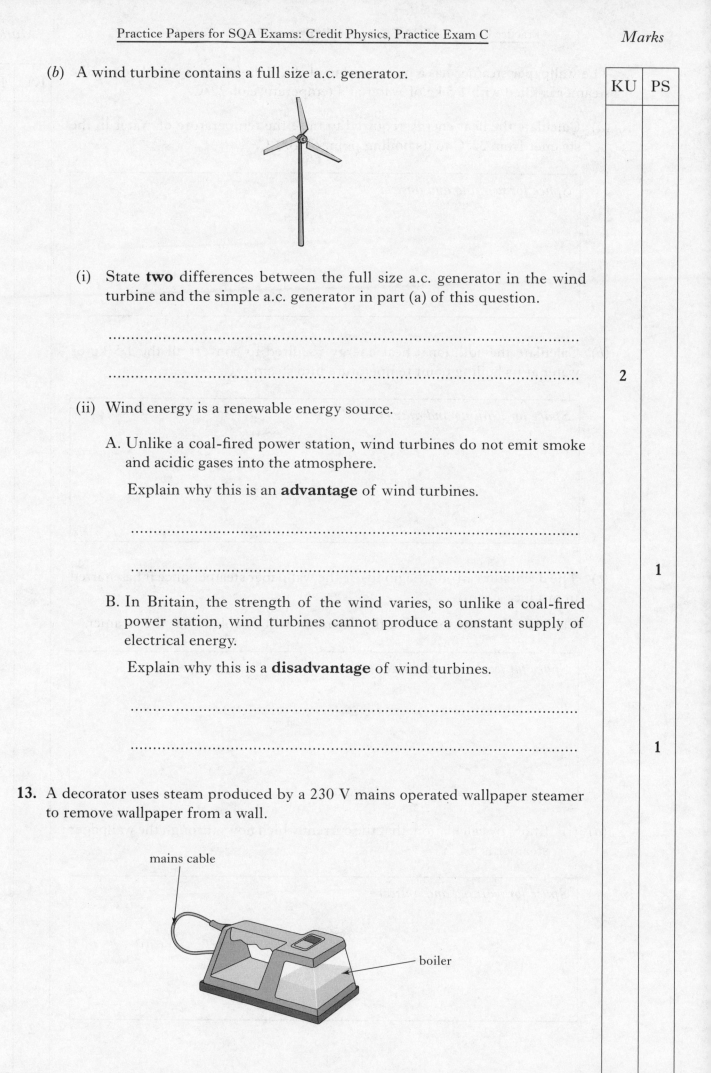

(i) State **two** differences between the full size a.c. generator in the wind turbine and the simple a.c. generator in part (a) of this question.

....

....

2

(ii) Wind energy is a renewable energy source.

A. Unlike a coal-fired power station, wind turbines do not emit smoke and acidic gases into the atmosphere.

Explain why this is an **advantage** of wind turbines.

....

....

1

B. In Britain, the strength of the wind varies, so unlike a coal-fired power station, wind turbines cannot produce a constant supply of electrical energy.

Explain why this is a **disadvantage** of wind turbines.

....

....

1

13. A decorator uses steam produced by a 230 V mains operated wallpaper steamer to remove wallpaper from a wall.

mains cable

boiler

The wallpaper steamer has a power rating of 2 kW. Before it is switched on, the steamer is filled with 1·5 kg of water at a temperature of 22°C

(a) Calculate the heat energy required to raise the temperature of water in the steamer from 22°C to its boiling point of 100°C.

> *Space for working and answer*

(b) Calculate the additional heat energy required to convert all the 1·5 kg of water at its boiling point temperature into steam.

> *Space for working and answer*

(c) The decorator can only begin to use the wallpaper steamer once it has started to produce steam.

Calculate the time for which the decorator will be able to use the steamer.

> *Space for working and answer*

(d) (i) Show, by calculation, that the current which flows through the wallpaper steamer is 8·7 A.

> *Space for working and answer*

KU	PS
	3
	3
2	
2	

(ii) Calculate the total quantity of charge which will flow through the wallpaper steamer when it is switched on for 25 minutes.

> *Space for working and answer*

14. The graph shows how the gravitational field strength of the earth changes as we travel further away from the surface of the earth.

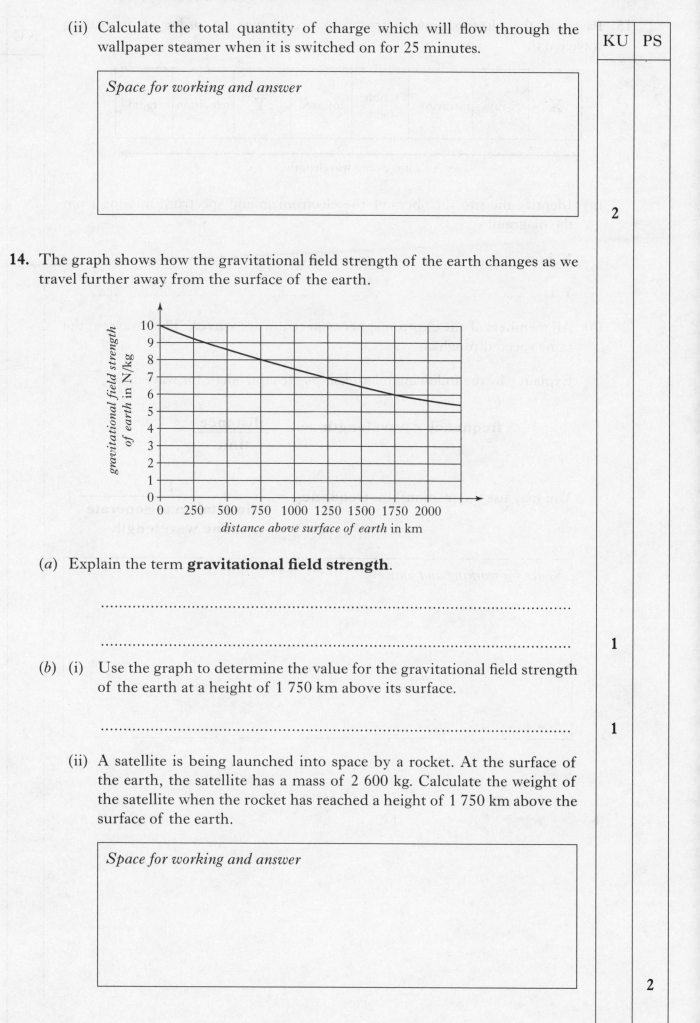

(*a*) Explain the term **gravitational field strength**.

....

....

(*b*) (i) Use the graph to determine the value for the gravitational field strength of the earth at a height of 1 750 km above its surface.

....

(ii) A satellite is being launched into space by a rocket. At the surface of the earth, the satellite has a mass of 2 600 kg. Calculate the weight of the satellite when the rocket has reached a height of 1 750 km above the surface of the earth.

> *Space for working and answer*

KU	PS
	2
2	
1	
	1
	2

15. Parts of the electromagnetic spectrum are shown, in order of increasing wavelength.

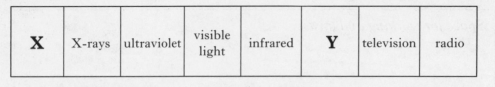

| X | X-rays | ultraviolet | visible light | infrared | Y | television | radio |

increasing wavelength

(*a*) Identify the two members of the electromagnetic spectrum missing from the diagram:

X is ....

Y is ....

(*b*) All members of the electromagnetic spectrum are **waves**. They travel at the same speed through air.

Explain why the following relationships are equivalent for waves:

$$\textbf{frequency} \times \textbf{wavelength} \quad \text{and} \quad \frac{\textbf{distance}}{\textbf{time}}$$

You may use the relationship: $\textbf{frequency} = \dfrac{1}{\textbf{time taken to generate one wavelength}}$

Space for working and answer

KU	PS
	2
	2

Credit Physics

| Practice Papers for SQA Exams | Time allowed: 1 hours, 45 minutes | Exam D |

Fill in these boxes and read what is printed below.

Full name of school

Town

Forename(s)

Surname

Read each question carefully.

Attempt **all** questions. You may answer them in any order.

Write your answers in the spaces provided.

Write as neatly as possible. If you decide to change your answer, cross it out neatly before rewriting it.

Answer in sentences wherever possible.

Any data needed will be on the data sheet at the back of the book.

Use appropriate numbers of significant figures in final answer.

Leckie×Leckie
Scotland's leading educational publishers

Marks

| | KU | PS |

1. (a) The Durris television transmitter, near Stonehaven, uses different UHF wavebands to transmit different television channels.

This information is shown in the table below.

Television channel	BBC1	BBC2	STV	Channel 4
UHF waveband (MHz)	479·25 to 487·25	527·25 to 535·25	503·25 to 511·25	559·25 to 567·25

(i) Explain why the television channels are transmitted using different wavebands.

....

1

(ii) Which television channel is transmitted at the **shortest** wavelength?

....

1

(b) Some television transmitters are fitted with **curved reflectors**.

Complete the diagram below to show how the curved reflector enables a beam of signals to be accurately **transmitted** to a far away receiver.

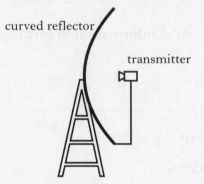

curved reflector

transmitter

(c) Use the terms **line build up, image retention** and **brightness variation** to describe how a moving picture is seen on the screen of a black and white television set.

....

....

....

....

....

2

3

2. The diagram shows two satellites orbiting the earth.

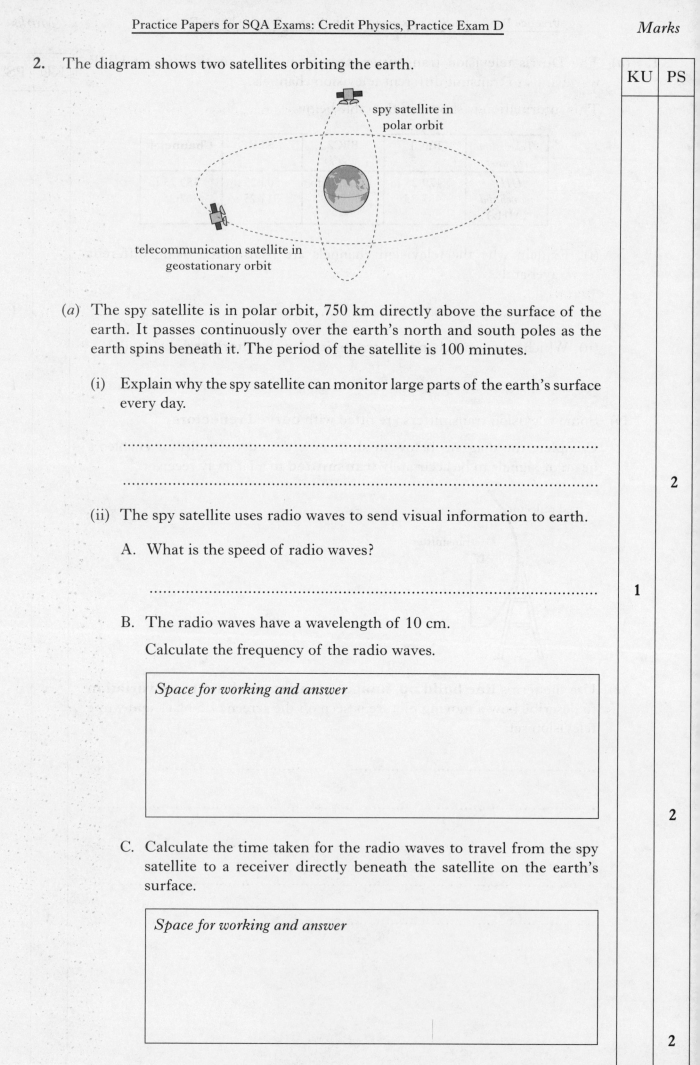

spy satellite in
polar orbit

telecommunication satellite in
geostationary orbit

(*a*) The spy satellite is in polar orbit, 750 km directly above the surface of the earth. It passes continuously over the earth's north and south poles as the earth spins beneath it. The period of the satellite is 100 minutes.

   (i) Explain why the spy satellite can monitor large parts of the earth's surface every day.

....

....

   (ii) The spy satellite uses radio waves to send visual information to earth.

     A. What is the speed of radio waves?

....

     B. The radio waves have a wavelength of 10 cm.

       Calculate the frequency of the radio waves.

Space for working and answer

     C. Calculate the time taken for the radio waves to travel from the spy satellite to a receiver directly beneath the satellite on the earth's surface.

Space for working and answer

KU	PS
	2
1	
	2
	2

(b) The telecommunication satellite is in a geostationary orbit, 36 000 km above the surface of the earth.

Suggest **two** reasons why the telecommunication satellite would **not** be of much use as a **spy satellite.**

....

....

2

3. (a) In Scotland, the **peak** voltage of the mains supply is 325 V.

How does the **peak** voltage of the mains supply compare with the value **usually** quoted?

....

1

(b) A simple d.c. electric motor is connected to a 6·0 V battery and a switch.

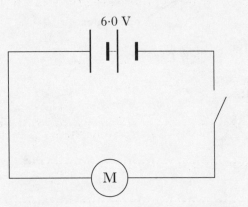

6·0 V

M

(i) When the switch is closed, a current of 48 mA flows through the motor.

For what time must the switch be **closed** in order for a total charge of 6·0 C to flow through the lamp?

┌───┐
│ *Space for working and answer*   │
│   │
│   │
│   │
│   │
└───┘

2

(ii) The 6·0 V battery is replaced with a 9·0 V battery.

How will this affect the quantity of energy given to the charges which will flow in the circuit when the switch is closed?

....

1

(iii) The diagram below shows the simple d.c. electric motor.

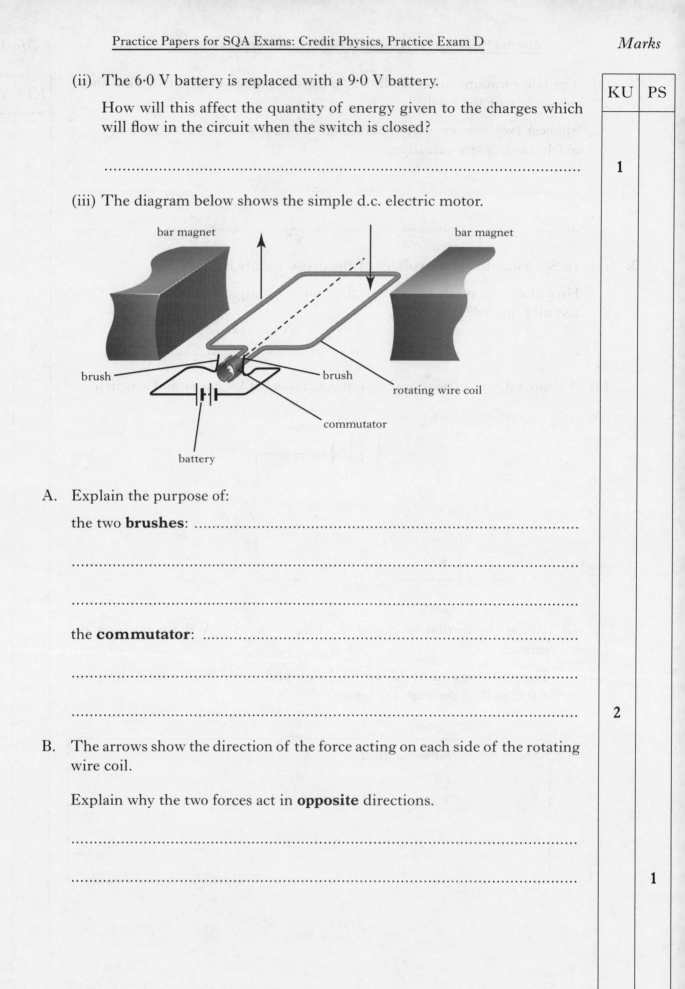

bar magnet

bar magnet

brush

brush

rotating wire coil

commutator

battery

A. Explain the purpose of:

the two **brushes**: ....

....

....

the **commutator**: ....

....

....

2

B. The arrows show the direction of the force acting on each side of the rotating wire coil.

Explain why the two forces act in **opposite** directions.

....

....

1

4. (a) A student connects three lamps to an Ohmmeter, as shown below. Each lamp has a resistance of $7 \cdot 2 \ \Omega$.

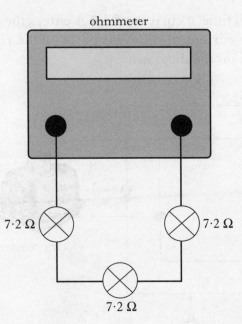

ohmmeter

$7 \cdot 2 \ \Omega$ $\quad$ $7 \cdot 2 \ \Omega$

$7 \cdot 2 \ \Omega$

(i) Calculate the value of the reading which will be displayed on the ohmmeter.

Space for working and answer

(ii) Explain why this method of connecting lamps would **not** be suitable for a household lighting circuit.

....

....

KU	PS
2	
	1

KU	PS

(b) The diagram below shows an electric toaster connected to a household ring main circuit. For simplicity, the earth wire has not been included.

At a particular instant in time, a current of 3·4 A enters the ring circuit. The direction of the electric current in different parts of the ring circuit at this instant in time is shown on the diagram.

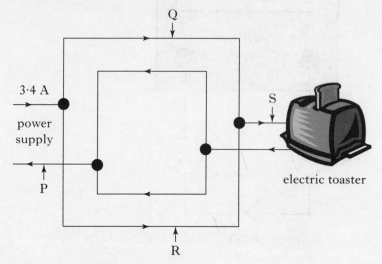

(i) Complete this table to show a possible value for the **size** of the electric current at each of the points **P**, **Q**, **R** and **S** in the ring main circuit.

Point in circuit	P	Q	R	S
Size of electric current at point in circuit (A)				

2

(ii) After 0·01 s, the current entering and leaving the ring circuit has **reversed** direction.

What does this indicate about the nature of the **power supply**?

....

1

(iii) A ring main circuit is the preferred method of wiring appliances in parallel for connection to the mains supply.

State **one** advantage a ring main circuit has over a standard parallel circuit.

....

....

1

(c) Use the relationship **P = IV** and the Ohm's law relationship **V = IR** to show that **electrical power** can be calculated using the relationship $P = I^2R$.

Space for working and answer

2

Marks

KU	PS

5. (a) A surgeon uses an endoscope to view inside a patient's body.

The endoscope consists of two bundles of optical fibres, side by side. Cold light is passed down optical fibre bundle X.

cold light passed down optical fibre bundle X

optical fibre bundle Y

doctor's eye

cold light source

(i) Name the process by which the cold light travels along an optical fibre.

....

1

(ii) Explain the purpose of optical fibre bundle Y.

....

....

1

(b) (i) The surgeon discovers a cancer tumour in the patient's stomach. She decides to destroy the cancer tumour using a laser beam and the endoscope.

Give **one** reason why a laser beam would be suitable for destroying the cancer tumour.

....

....

1

(ii) Describe how the surgeon could use the laser beam and endoscope to destroy the cancer tumour.

....

....

....

2

(c) Before operating on the patient, the surgeon requests an X-ray image of the patient's stomach.

Explain why an X-ray image obtained by **computerised tomography** might be more appropriate.

....

....

1

6. (*a*) A doctor suspects one of a patient's two kidneys is not working normally.

To investigate, the doctor inserts a radioactive liquid tracer into both of the patient's kidneys. The kidneys should pass the radioactive liquid tracer quickly to the patient's bladder.

The process is observed using a gamma camera.

kidneys

bladder

(i) The radioactive liquid tracer emits gamma radiation.

Suggest **two** reasons why gamma radiation is preferred to alpha radiation for the liquid tracer.

....

....

(ii) The graph shows how the activity detected from each of the patient's kidneys due to the radioactive liquid tracer they contain changes with time.

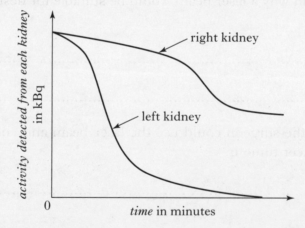

State and explain which of the patient's kidneys is **not** working normally.

....

....

....

(iii) The observation of the patient's kidneys was carried out for 25 minutes. Four radioactive liquid tracers (A, B, C and D), with the following **halflife** values, were available for use:

A = 2 minutes   B = 10 minutes   C = 4 hours   D = 2 days

KU	PS
	2
	2

Marks

KU	PS

State and explain which tracer would be most suitable for inserting into the patient's kidneys.

....

....

....

2 (PS)

(b) (i) For living materials, the biological effect of radiation depends on **two** factors.

State these **two** factors.

....

....

2 (KU)

(ii) State the unit for **dose equivalent**.

....

1 (KU)

(c) Radiation can be detected using a **Geiger-Müller tube**, **film badge** or **scintillation counter**.

Select **one** of these devices and describe how it detects radiation.

Device selected: ....

How it detects radiation: ....

....

....

2 (KU)

7. (a) A vending machine is fitted with an **alarm system**.

The **logic diagram** for the **alarm system** is shown below.

- If the door is opened, the logic level at A changes from **0** to **1**.
- If someone breaks the lock, the logic level at B changes from **1** to **0**.

door switch │ A

warning LED and buzzer │ D

Y

contact switch in lock │ B   X   C

Marks

KU	PS

(i) Explain why **logic gate X** is necessary in the circuit.

....

....

1

(ii) The inputs to **logic gate Y** are **A** and **C**. The output is **D**. Draw the truth table for **logic gate Y**.

Space for answer

2

(b) The circuit diagram for the warning LED is shown.

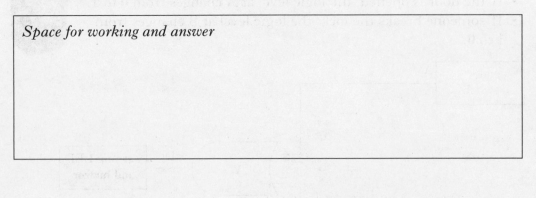

When the LED is lit and operating correctly, the voltage across the LED is 2·2 V and the current passing through it is 10 mA.

Calculate the resistance of protective resistor R.

Space for working and answer

3

8. This circuit can be used to switch on a warning buzzer in a greenhouse when the light level becomes too high.

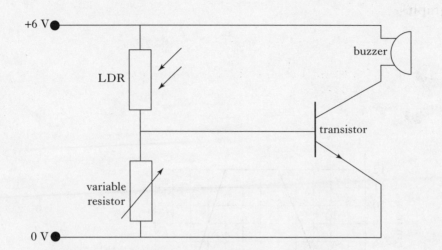

(a) The resistance of the variable resistor is set at 2 kΩ.

At a particular light level, the resistance of the LDR is 18 kΩ.

Calculate the voltage across the LDR at this light level.

> *Space for working and answer*

2

(b) The transistor starts to conduct when the voltage across its input is 0·7 V.

Show, by calculating the voltage across the variable resistor, that the buzzer will **not** be switched on at this light level.

> *Space for working and answer*

1

(c) What adjustment could be made to the resistance setting of the variable resistor in order for the buzzer to be switched on at this light level?

....

....

1

	KU	PS

9. An abseiler has a mass of 68 kg.

The speed of the abseiler as he slides down a rope for 6·0 s is shown on the speed-time graph.

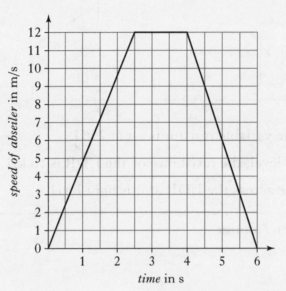

time in s

(*a*) Calculate the distance the abseiler slides down the rope during the 6·0 s.

> *Space for working and answer*

(*b*) (i) Calculate the **average speed** of the abseiler as he slides down the rope.

> *Space for working and answer*

2

2

	KU	PS

(ii) Explain why the **instantaneous speed** of the abseiler as he slides down the rope is not always the same as his **average speed**.

....

....

(c) (i) Show, by calculation, that during part of his descent, the abseiler **decelerates** at 6·0 m/s².

> *Space for working and answer*

(ii) Calculate the size of the unbalanced force required to produce this deceleration.

> *Space for working and answer*

10. The **total distance** a moving car takes to stop is made up of two parts:

- **thinking distance** – the distance the car travels between the time the driver sees an object in road and the time the driver presses the brake pedal

- **braking distance** – the distance the car takes to stop after the brake pedal has been pressed

The table below shows the **thinking distance** and the **stopping distance** of a typical family car travelling at various speeds.

Speed of car (miles per hour)	Thinking distance (m)	Braking distance (m)	Total distance car takes to stop (m)
30	9	14	23
50	15	38	53
60	18	54	72
70	21	75	96

(*a*) Why is it **not** possible for a car driver to press the brake pedal **immediately** when he or she sees an object in the road?

....

....

1

(*b*) It is assumed that, no matter how fast a car is travelling, the driver always takes the same time to press the brake pedal after seeing an object in the road.

Explain why the **thinking distance** increases as the speed of the car increases.

....

....

1

(*c*) The **braking distance** of a car increases as the **speed** (and therefore the **kinetic energy**) of the car increases.

Explain this in terms of the **work done** by the car **brakes** (which exert a constant braking force, no matter the speed of the car) and the **time** taken to bring the car to rest.

....

....

....

2

11. A nuclear power station produces electricity for the National Grid.

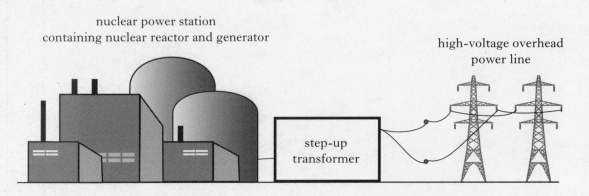

nuclear power station
containing nuclear reactor and generator

high-voltage overhead
power line

step-up
transformer

(a) In the nuclear reactor of the power station, a neutron strikes the nucleus of a uranium atom. The uranium nucleus splits into two smaller parts and three neutrons are released.

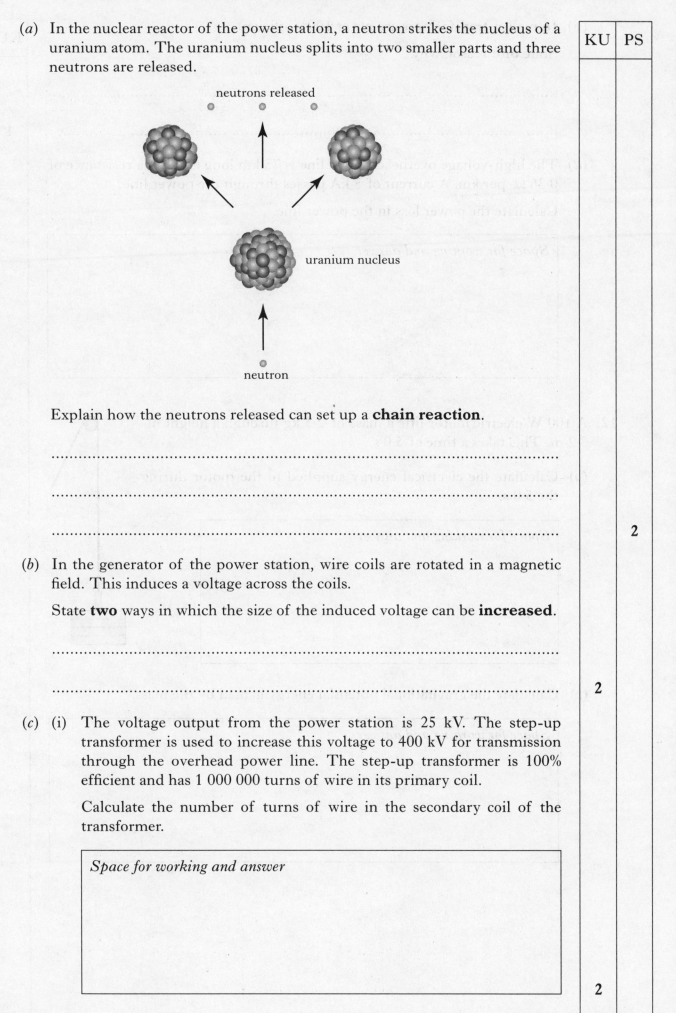

neutrons released

uranium nucleus

neutron

Explain how the neutrons released can set up a **chain reaction**.

....

....

....

KU	PS
2	

(b) In the generator of the power station, wire coils are rotated in a magnetic field. This induces a voltage across the coils.

State **two** ways in which the size of the induced voltage can be **increased**.

....

....

2

(c) (i) The voltage output from the power station is 25 kV. The step-up transformer is used to increase this voltage to 400 kV for transmission through the overhead power line. The step-up transformer is 100% efficient and has 1 000 000 turns of wire in its primary coil.

Calculate the number of turns of wire in the secondary coil of the transformer.

Space for working and answer

KU	PS
2	

	KU	PS

(ii) In reality, transformers are never 100% efficient.

State **one** reason why.

....

....

1

(d) The high-voltage overhead power line is 75 km long and has a resistance of 0·36 Ω per km. A current of 5 kA passes through the power line.

Calculate the power loss in the power line.

> *Space for working and answer*

3

12. A 100 W electric motor lifts a mass of 2·5 kg through a height of 3·2 m. This takes a time of 5·0 s.

(a) Calculate the electrical energy supplied to the motor during the 5·0 s.

> *Space for working and answer*

2

(b) Calculate the gravitational potential energy gained by the mass.

> *Space for working and answer*

2

(c) Calculate the percentage efficiency of the electric motor during the lifting process.

> *Space for working and answer*

2

13. (a) At the instant it is launched from the earth's surface, the mass of a space rocket is 1 500 000 kg (1.5×10^6 kg).

   (i) Calculate the weight of the space rocket on earth at the instant it is launched.

> *Space for working and answer*

2

   (ii) As the space rocket approaches **outer space**, its weight decreases. Suggest **one** reason why.

   ....

1

(b) At the instant of launch, no air resistance forces act on the space rocket.

On the diagram below, *label* and show the *direction* of **two forces** which do act on the space rocket at the instant of launch.

2

14. A meteorite of mass 120 kg enters the earth's atmosphere with a speed of 2 000 m/s. Due to friction between the atmosphere and the meteorite, the speed of the meteorite decreases to 900 m/s.

KU	PS

(a) Show, by calculation, that the **decrease** in kinetic energy of the meteorite is 191 400 000 J.

> *Space for working and answer*

3

(b) The material from which the meteorite is made has a specific heat capacity of 1 100 J/kg °C.

Calculate the **rise in temperature** of the meteorite on entering the earth's atmosphere.

> *Space for working and answer*

3

(c) State **one** reason why the increase in temperature of the meteorite may be different to the value calculated in part (b).

....

.... 1

1. (a) 625×25 **(1/2 mark)**

        $= 15\ 625$ lines   **(1/2 mark)**

> *HINT*   This part of the question involves a simple calculation — hence its 1 mark value.

(b)   The eye/brain system 'holds' each picture for a fraction of a second (image retention). **(1 mark)**

      The eye/brain system blends successive pictures together.   **(1 mark)**

> *HINT*   You must be able to explain **image retention** and the ability of the eye/brain to blend successive pictures together to create the illusion of a continuous picture.

(c)   The electron gun fires more electrons at the screen when the electron beam is hitting that part of the screen.   **(1 mark)**

> *HINT*   You must know that if more electrons hit a point on the television screen in a given time, the brightness of that point on the screen will increase.

2. (a) (i)   diffraction   **(1 mark)**

> *HINT*   You must know the meaning of **diffraction**.
>
> You must be able to spell the word correctly — you will not be awarded a mark for an incorrect spelling.

    (ii)

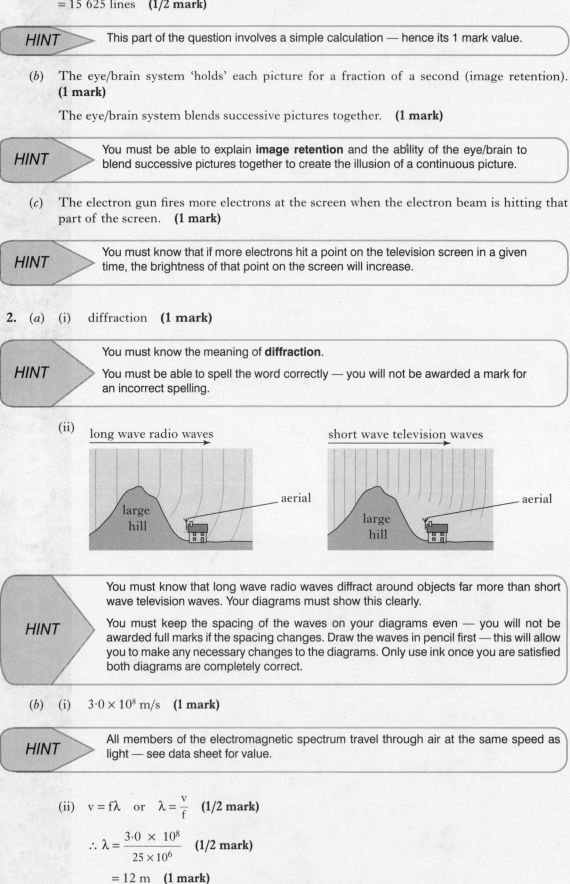

> *HINT*   You must know that long wave radio waves diffract around objects far more than short wave television waves. Your diagrams must show this clearly.
>
> You must keep the spacing of the waves on your diagrams even — you will not be awarded full marks if the spacing changes. Draw the waves in pencil first — this will allow you to make any necessary changes to the diagrams. Only use ink once you are satisfied both diagrams are completely correct.

(b) (i) $3 \cdot 0 \times 10^8$ m/s   **(1 mark)**

> *HINT*   All members of the electromagnetic spectrum travel through air at the same speed as light — see data sheet for value.

    (ii) $v = f\lambda$ or $\lambda = \dfrac{v}{f}$ **(1/2 mark)**

        $\therefore \lambda = \dfrac{3 \cdot 0 \times 10^8}{25 \times 10^6}$ **(1/2 mark)**

          $= 12$ m   **(1 mark)**

> **HINT** Check the multiplication factor on the data sheet.

(*c*)   range increases   **(1 mark)**

Transmitted (incident) signal travels further to right before undergoing reflection off Heaviside layer.   **(1 mark)**

> **HINT** Think carefully. To start to answer this, you could draw the Heaviside layer slightly further away from the earth's surface and continue the incident ray from the transmitter until it hits the layer. Then you could draw the new reflected ray and observe where it would hit the earth's surface — you could extend the curve of the earth on the diagram. Use a ruler and a pencil in case you need to make any changes to your diagram.
>
> HOWEVER YOU DO NOT NEED TO DRAW ANYTHING ON THE DIAGRAM AND NO MARKS WILL BE AWARDED FOR DRAWING ON THE DIAGRAM — YOU MUST PROVIDE A WRITTEN EXPLANATION TO BE AWARDED ANY MARKS.

3.   (*a*) $P = \dfrac{V^2}{R}$ **(1/2 mark)**

$= \dfrac{12^2}{60}$ **(1/2 mark)**

$= 2 \cdot 4$ W   **(1 mark)**

> **HINT** You must understand that this is a parallel circuit, so the voltage across each lamp is 12 V (the supply voltage).

(*b*) $\dfrac{1}{R_T} = \dfrac{1}{R_1} + \dfrac{1}{R_2} + \dfrac{1}{R_3}$ **(1/2 mark)**

$= \dfrac{1}{120} + \dfrac{1}{60} + \dfrac{1}{60}$ **(1/2 mark)**

$= \dfrac{5}{120}$

$\therefore R_T = \dfrac{120}{5}$

$= 24 \ \Omega$ **(1 mark)**

> **HINT** You must understand that the components are connected in parallel.
>
> You must be able to calculate the total resistance of two or more circuit components connected in parallel.

(*c*) $Q = It$ **(1/2 mark)**

$= 500 \times 10^{-3} \times (3 \times 60)$ **(1/2 mark)**

$= 90$ C   **(1 mark)**

> **HINT** You must convert 500 mA to A. (Check the multiplication factor on the data sheet.)
>
> You must also convert 3 minutes to seconds. (Multiply by 60).
>
> The values used in the calculation must be in A and s.

(*d*)   (i)   Light emitting diodes require (much) less current than lamps to light.
or   Light emitting diodes emit (much) less heat energy than lamps/transfer (much) less electrical energy to heat energy than lamps.   **(1 mark)**

> **HINT** You must know some of the advantages an LED has over a lamp.

(ii)

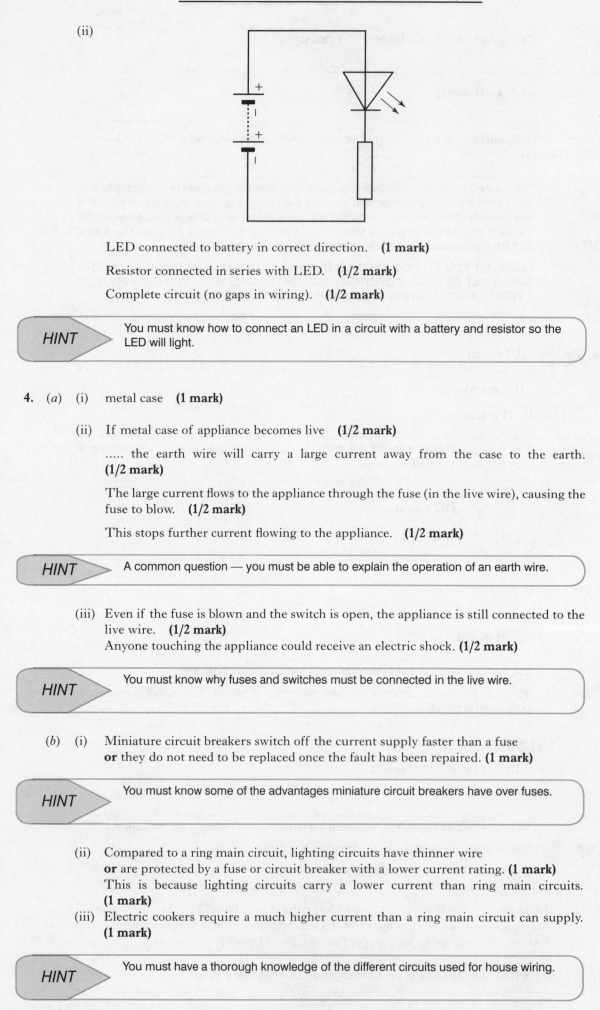

LED connected to battery in correct direction.   **(1 mark)**

Resistor connected in series with LED.   **(1/2 mark)**

Complete circuit (no gaps in wiring).   **(1/2 mark)**

> **HINT** You must know how to connect an LED in a circuit with a battery and resistor so the LED will light.

4. (a) (i)   metal case   **(1 mark)**

(ii) If metal case of appliance becomes live   **(1/2 mark)**

..... the earth wire will carry a large current away from the case to the earth. **(1/2 mark)**

The large current flows to the appliance through the fuse (in the live wire), causing the fuse to blow.   **(1/2 mark)**

This stops further current flowing to the appliance.   **(1/2 mark)**

> **HINT** A common question — you must be able to explain the operation of an earth wire.

(iii) Even if the fuse is blown and the switch is open, the appliance is still connected to the live wire.   **(1/2 mark)**
Anyone touching the appliance could receive an electric shock. **(1/2 mark)**

> **HINT** You must know why fuses and switches must be connected in the live wire.

(b) (i)   Miniature circuit breakers switch off the current supply faster than a fuse **or** they do not need to be replaced once the fault has been repaired. **(1 mark)**

> **HINT** You must know some of the advantages miniature circuit breakers have over fuses.

(ii) Compared to a ring main circuit, lighting circuits have thinner wire **or** are protected by a fuse or circuit breaker with a lower current rating. **(1 mark)**
This is because lighting circuits carry a lower current than ring main circuits. **(1 mark)**

(iii) Electric cookers require a much higher current than a ring main circuit can supply. **(1 mark)**

> **HINT** You must have a thorough knowledge of the different circuits used for house wiring.

5. (*a*) (i) Time taken for the activity of a radioactive substance to halve.
or Time taken for half of the atoms present in a radioactive substance to decay.
(1 mark)

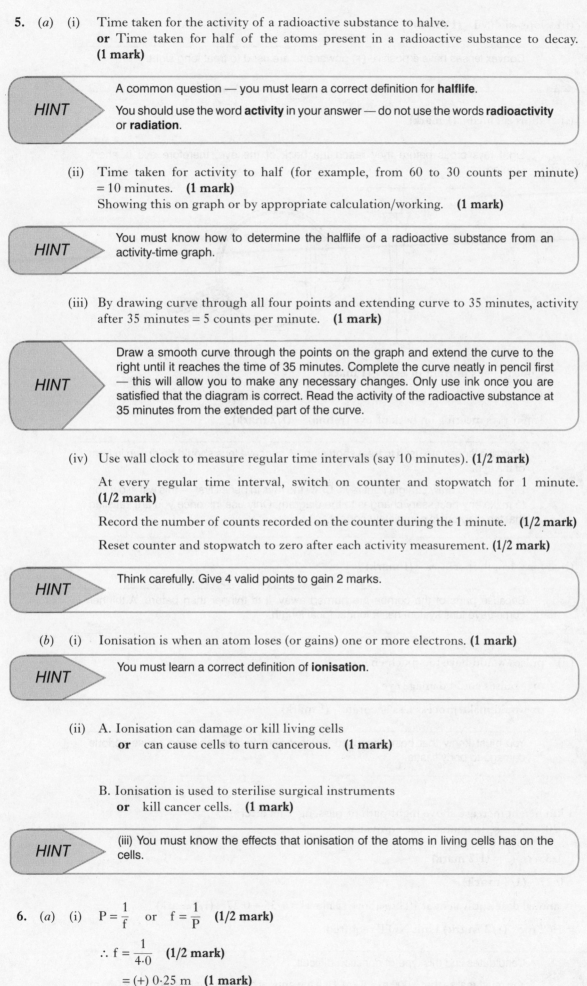

> **HINT**
>
> A common question — you must learn a correct definition for **halflife**.
>
> You should use the word **activity** in your answer — do not use the words **radioactivity** or **radiation**.

(ii) Time taken for activity to half (for example, from 60 to 30 counts per minute)
= 10 minutes.   **(1 mark)**
Showing this on graph or by appropriate calculation/working.   **(1 mark)**

> **HINT**
>
> You must know how to determine the halflife of a radioactive substance from an activity-time graph.

(iii) By drawing curve through all four points and extending curve to 35 minutes, activity
after 35 minutes = 5 counts per minute.   **(1 mark)**

> **HINT**
>
> Draw a smooth curve through the points on the graph and extend the curve to the right until it reaches the time of 35 minutes. Complete the curve neatly in pencil first — this will allow you to make any necessary changes. Only use ink once you are satisfied that the diagram is correct. Read the activity of the radioactive substance at 35 minutes from the extended part of the curve.

(iv) Use wall clock to measure regular time intervals (say 10 minutes). **(1/2 mark)**

At every regular time interval, switch on counter and stopwatch for 1 minute.
(1/2 mark)

Record the number of counts recorded on the counter during the 1 minute.   **(1/2 mark)**

Reset counter and stopwatch to zero after each activity measurement. **(1/2 mark)**

> **HINT**
>
> Think carefully. Give 4 valid points to gain 2 marks.

(*b*) (i) Ionisation is when an atom loses (or gains) one or more electrons. **(1 mark)**

> **HINT**
>
> You must learn a correct definition of **ionisation**.

(ii) A. Ionisation can damage or kill living cells
or   can cause cells to turn cancerous.   **(1 mark)**

B. Ionisation is used to sterilise surgical instruments
or   kill cancer cells.   **(1 mark)**

> **HINT**
>
> (iii) You must know the effects that ionisation of the atoms in living cells has on the cells.

6. (*a*) (i) $P = \dfrac{1}{f}$ or $f = \dfrac{1}{P}$ **(1/2 mark)**

∴ $f = \dfrac{1}{4 \cdot 0}$ **(1/2 mark)**

= (+) 0·25 m   **(1 mark)**

(ii) long-sighted **(1 mark)**

> *HINT* — Convex lenses have a positive (+) power and are used to treat long sight.

(b) (i) short-sighted **(1 mark)**

> *HINT* — Light rays cross before they reach the back of the eye, therefore eye is short-sighted.

(ii)

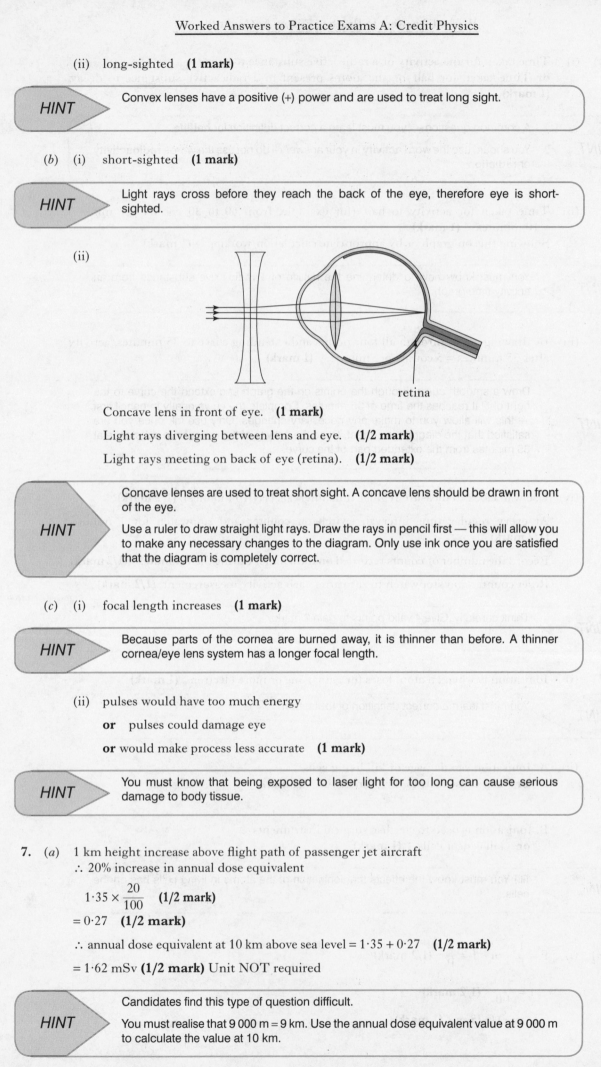

retina

Concave lens in front of eye. **(1 mark)**

Light rays diverging between lens and eye. **(1/2 mark)**

Light rays meeting on back of eye (retina). **(1/2 mark)**

> *HINT* — Concave lenses are used to treat short sight. A concave lens should be drawn in front of the eye.
>
> Use a ruler to draw straight light rays. Draw the rays in pencil first — this will allow you to make any necessary changes to the diagram. Only use ink once you are satisfied that the diagram is completely correct.

(c) (i) focal length increases **(1 mark)**

> *HINT* — Because parts of the cornea are burned away, it is thinner than before. A thinner cornea/eye lens system has a longer focal length.

(ii) pulses would have too much energy

or pulses could damage eye

or would make process less accurate **(1 mark)**

> *HINT* — You must know that being exposed to laser light for too long can cause serious damage to body tissue.

7. (a) 1 km height increase above flight path of passenger jet aircraft

∴ 20% increase in annual dose equivalent

$$1·35 \times \frac{20}{100} \quad \textbf{(1/2 mark)}$$

$$= 0·27 \quad \textbf{(1/2 mark)}$$

∴ annual dose equivalent at 10 km above sea level $= 1·35 + 0·27$ **(1/2 mark)**

$= 1·62$ mSv **(1/2 mark)** Unit NOT required

> *HINT* — Candidates find this type of question difficult.
>
> You must realise that 9 000 m = 9 km. Use the annual dose equivalent value at 9 000 m to calculate the value at 10 km.

(b) Because of the time they spend flying, pilots are exposed to greater levels of cosmic radiation than members of the general public. **(1/2 mark)**

Once they have received a maximum dose, they must stop flying so as not to risk their health. **(1/2 mark)**

HINT
You must understand that limits are set for the maximum safe value of exposure a human being can have to different types of radiation. After a certain time, when the maximum safe exposure value for a type of radiation has been reached, a person is not allowed further exposure to that type of radiation for the remainder of the year.

8. (a) (i)   NOT gate or inverter   **(1 mark)**

   (ii)   AND gate   **(1/2 mark)**

HINT
You must be able to identify the symbols for AND, OR and NOT (inverter) logic gates.

(iii)

U	V	W	X	Y	Z
0	0	1	0	0	0
0	1	1	1	0	0
1	0	0	0	0	0
1	1	0	0	0	0
0	0	1	0	1	0
0	1	1	1	1	1
1	0	0	0	1	0
1	1	0	0	1	0

All of column W correct.   **(1 mark)**

All of column X correct.   **(1 mark)**

All of column Z correct.   **(1 mark)**

HINT
Complete the truth table in pencil first — this will allow you to make any necessary changes. Only use ink once you are satisfied that all the entries in the table are correct.

(b) (i)   Remove logic gate R   **(1 mark)**

HINT
You should be able to suggest appropriate adjustments to logic circuits.

   (ii)   LDR or solar cell   **(1 mark)**

HINT
You must be able to select a suitable input device for a given application.

9. (a)   buzzer   **(1 mark)**

HINT
You must be able to select a suitable output device for a given application.

(b) (i) $5 - 0.7$ **(1/2 mark)**
            $= 4.3$ V **(1/2 mark)**

(ii) $\dfrac{V_1}{V_2} = \dfrac{R_1}{R_2}$ **(1/2 mark)**

$\therefore \dfrac{4.3}{0.7} = \dfrac{15}{R_{thermistor}}$ **(1/2 mark)**

$\therefore R_{thermistor} = 2.4\ k\Omega$ **(1 mark)**

(c) To set the temperature at which the transistor switches the warning output device on. **(1 mark)**

(d) As temperature falls, resistance of thermistor increases **(1/2 mark)**

$\therefore$ Voltage across thermistor (and voltage across base of transistor) increases. **(1/2 mark)**

When voltage across base of transistor reaches 0.7 V, transistor starts to conduct. **(1/2 mark)**

$\therefore$ Output device is switched on (by transistor). **(1/2 mark)**

10. (a) (i) $50 - 25$ **(1/2 mark)**
                  $= 25$ km/h **(1/2 mark)**

(ii) $a = \dfrac{v - u}{t}$ **(1/2 mark)**

$\therefore a = \dfrac{70 - 50}{36}$ **(1/2 mark)**

$= 0.56$ km/h/s **(1 mark)**

(b) Unbalanced force (F) = 266·6 – 255·4   **(1/2 mark)**

$$= 11·2 \text{ N} \quad \textbf{(1/2 mark)}$$

$$F = ma \text{ or } a = \frac{F}{m} \quad \textbf{(1/2 mark)}$$

$$\therefore a = \frac{11·2}{68·0} \quad \textbf{(1/2 mark)}$$

$$= 0·16 \text{ m/s}^2 \quad \textbf{(1 mark)}$$

HINT Be careful of the change in acceleration units from km/h/s to m/s².
This part of the question is worth 3 marks, which indicates more than one part to the calculation. First, calculate the size of the unbalanced force acting on the cyclist, then calculate the value for their acceleration using the unbalanced force and mass values.

11. (a) $E_k = \frac{1}{2}mv^2$ or $v = \sqrt{\dfrac{E_k}{1/2\ m}}$ **(1/2 mark)**

$$\therefore v = \sqrt{\frac{175}{0·5 \times 0·45}} \quad \textbf{(1/2 mark)}$$

$$= 28 \text{ m/s} \quad \textbf{(1 mark)}$$

HINT The first step is to use the relationship $E_k = 1/2\ mv^2$ to calculate v².
You must then take the square root of this answer to obtain the value for v.

(b) (i)   175 J   **(1 mark)**

HINT You must understand that:
work done in bringing football to rest = decrease in kinetic energy of football to zero

(ii) $E_w = Fd$ or $F = \dfrac{E_w}{d}$ **(1/2 mark)**

$$\therefore F = \frac{175}{0·125} \quad \textbf{(1/2 mark)}$$

$$= 1\,400 \text{ N} \quad \textbf{(1 mark)}$$

HINT You must convert 12·5 cm to m. (Divide by 100).
The distance value used in the calculation must be in m.

12. (a) (i)   The power station can provide electrical energy for short periods of time when the demand is high.   **(1 mark)**

HINT You must know the advantages and disadvantages of several energy sources.

(ii) Agricultural land may be lost.
or People may lose their home due to the flooding.   **(1 mark)**

(b) $150\,000 \times 28$ **(1/2 mark)**

$$= 4\,200\,000 \text{ MJ (or } 4·2 \times 10^6 \text{ MJ)} \quad \textbf{(1/2 mark)}$$

HINT This part of the question involves a simple calculation, hence its 1 mark value.

13. (a) The control apparatus allows us to measure the mass of ice which melts due to the heat energy supplied by the surroundings — not by the immersion heater. **(1 mark)**

We can therefore calculate the actual mass of ice which melts due to the heat energy supplied by the immersion heater. **(1 mark)**

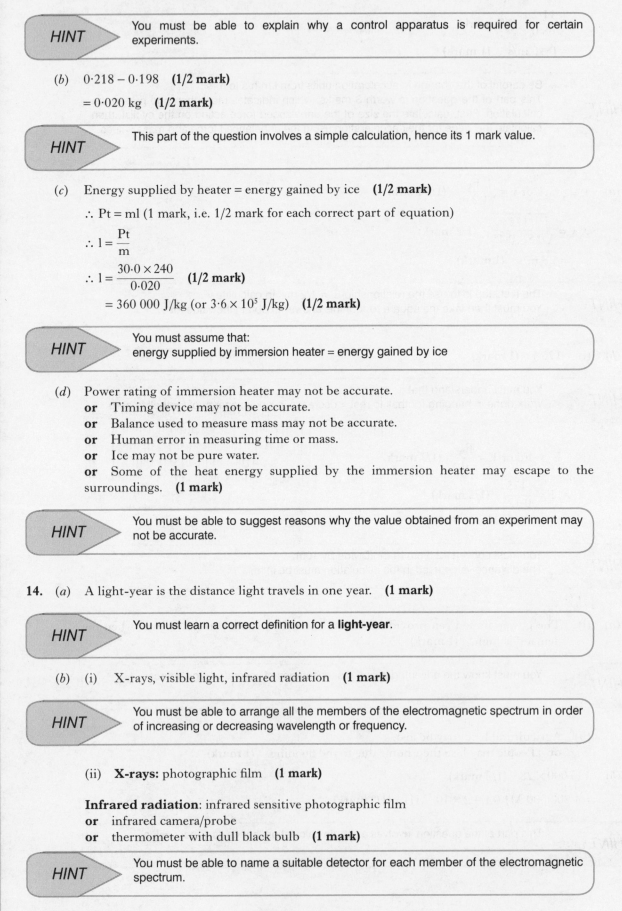

HINT > You must be able to explain why a control apparatus is required for certain experiments.

(b) $0.218 - 0.198$ **(1/2 mark)**

$= 0.020$ kg **(1/2 mark)**

HINT > This part of the question involves a simple calculation, hence its 1 mark value.

(c) Energy supplied by heater = energy gained by ice **(1/2 mark)**

∴ Pt = ml (1 mark, i.e. 1/2 mark for each correct part of equation)

∴ $l = \dfrac{Pt}{m}$

∴ $l = \dfrac{30 \cdot 0 \times 240}{0 \cdot 020}$ **(1/2 mark)**

$= 360\ 000$ J/kg (or $3 \cdot 6 \times 10^5$ J/kg) **(1/2 mark)**

HINT > You must assume that:
energy supplied by immersion heater = energy gained by ice

(d) Power rating of immersion heater may not be accurate.
or Timing device may not be accurate.
or Balance used to measure mass may not be accurate.
or Human error in measuring time or mass.
or Ice may not be pure water.
or Some of the heat energy supplied by the immersion heater may escape to the surroundings. **(1 mark)**

HINT > You must be able to suggest reasons why the value obtained from an experiment may not be accurate.

14. (a) A light-year is the distance light travels in one year. **(1 mark)**

HINT > You must learn a correct definition for a **light-year**.

(b) (i) X-rays, visible light, infrared radiation **(1 mark)**

HINT > You must be able to arrange all the members of the electromagnetic spectrum in order of increasing or decreasing wavelength or frequency.

(ii) **X-rays**: photographic film **(1 mark)**

Infrared radiation: infrared sensitive photographic film
or infrared camera/probe
or thermometer with dull black bulb **(1 mark)**

HINT > You must be able to name a suitable detector for each member of the electromagnetic spectrum.

(c) (i) Image formed by telescope would be brighter. **(1 mark)**
This is because more light would be collected by the larger diameter objective lens. **(1 mark)**

> **HINT** You must be able to explain how changing the diameter of the objective lens of an optical telescope affects the brightness of the image it forms.

(ii) A. Light pollution from street lamps, car headlights, etc. **or** Air pollution due to traffic, factories, etc. **(1 mark)**

B. Reduce light emitted by street lamps, etc.
or implement air pollution control measures
or move observatory to darker location, e.g. out of city
or take observation from satellite or space craft above the earth's atmosphere.
(1 mark)

> **HINT** You should think of typical conditions in a city at night time and how their effect on the image formed by an optical telescope can be minimised.

15. (a) (i) Volleyball exerts force on hand. **(1 mark)**

> **HINT** You must understand Newton's third law and understand Newton pairs.

(ii) The forces act on different objects (the volleyball and the hand). **(1 mark)**

> **HINT** You must know that:
> • Newton pairs are equal and opposite forces which act on different objects.
> • Balanced forces are equal and opposite forces which act on the same object.

(b) (i) Motion is a combination of constant speed in horizontal direction **(1 mark)**
and downward acceleration (due to gravity) in vertical direction. **(1 mark)**

> **HINT** You must be able to explain projectile motion.

(ii) Horizontal motion: $d = vt$ or $t = \dfrac{d}{v}$ **(1/2 mark)**

$\therefore t = \dfrac{6\cdot 3}{8\cdot 4}$ **(1/2 mark)**

$= 0\cdot 75$ s **(1 mark)**

> **HINT** You must consider the horizontal motion of the volleyball.
> The horizontal speed remains constant throughout its motion.
> The relationship **d (horizontal) = v (horizontal) × t** applies.

(iii) Vertical motion: $a = \dfrac{v-u}{t}$ **(1/2 mark)**

$\therefore 10 = \dfrac{v-u}{0\cdot 75}$ **(1/2 mark)**

$10 = \dfrac{v-0}{0\cdot 75}$

$\therefore v = 10 \times 0\cdot 75$

$= 7\cdot 5$ m/s **(1 mark)**

> **HINT** You must consider the downward vertical motion of the volleyball.
> The relationship $a = \dfrac{v-u}{t}$ applies.
> (a is the downward acceleration of the volleyball due to gravity = 10 m/s².)

1. (*a*) (i)   ELF waves do not reflect off the surface of seas/oceans.   **(1 mark)**

(ii)   LF radio waves have a longer wavelength than UHF television waves.   **(1 mark)**
Therefore, greater diffraction around hills with LF radio waves.   **(1 mark)**

Make sure you understand **diffraction** — the ability of waves to travel around solid obstacles.

Long wave (low frequency) waves diffract more than short wave (high frequency) waves.

To obtain full marks, your answer must include the word **diffraction** or **diffract** — **bending** is not a suitable description and will gain no marks. You will not be awarded marks if you spell **diffraction** or **diffract** incorrectly.

(*b*) (i)

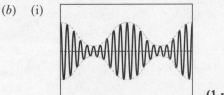

(1 mark)

HINT   Complete the diagram neatly in pencil first — this will allow you to make any necessary changes. Only use ink once you are satisfied that the diagram is correct.

(ii)   amplitude modulation   **(1 mark)**

HINT   The amplitude of the radio carrier wave is modulated (changed), hence the term **amplitude modulation**.

2. (*a*) (i)   Better signal quality: optical fibres   **(1/2 mark)**
Highest signal capacity: optical fibres   **(1/2 mark)**
Highest signal speed: metal wires   **(1/2 mark)**
Lowest cost: optical fibres   **(1/2 mark)**

HINT   The only advantages metal wires have over optical fibres are:

• higher signal speed in metal wires;

• metal wires are easier to join.

(ii)   A.

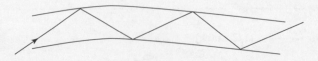

• straight lines/ruler used   **(1/2 mark)**
• all pairs of angles of incidence and reflection look equal   **(1/2 mark)**
• no part of light ray outside optical fibre/all reflections take place at walls of fibre **(1/2 mark)**
• no more than 5 reflections   **(1/2 mark)**

HINT

> This is a common question, either for Telecommunication (signal transfer) or Health Physics (the endoscope).
>
> Light travels in straight lines — you must use a ruler.
>
> Complete the diagram neatly in pencil first — this will allow you to make any necessary changes. Only use ink once you are satisfied the diagram is correct.
>
> All reflections must take place at the inside walls of the optical fibre.
>
> All pairs of angles of incidence and reflection must look equal.
>
> You will lose 1 mark if you show more than 5 reflections.

B. total internal reflection **(1 mark)**

HINT

> If you do not include the word **total**, you will not be awarded the mark.

(iii) A. Trace Y **(1/2 mark)**
because it has the higher amplitude **(1/2 mark)**

HINT

> Higher amplitude = louder

B. Trace Y **(1/2 mark)**
because it has the shorter wavelength **(1/2 mark)**

HINT

> Higher frequency = greater number of wavelengths

(b) (i) the signal can be transmitted in one direction/directed towards the receiver
or the strength of the signal is increased **(1 mark)**

HINT

> You must know the advantages of fitting curved reflector dishes to transmitter or receiver aerials.

(ii) A. $3 \cdot 0 \times 10^8$ m/s (or 300 000 000 m/s) **(1 mark)**

HINT

> All members of the electromagnetic spectrum travel through air at the same speed as light — see data sheet for value.

B. $v = f\lambda$ or $\lambda = \dfrac{v}{f}$ **(1/2 mark)**

$\therefore \lambda = \dfrac{3 \cdot 0 \times 10^8}{1\,500 \times 10^6}$ **(1/2 mark)**

$\therefore \lambda = 0 \cdot 2$ m **(1 mark)**

HINT

> Check the multiplication factor on the data sheet.

3. (a) mean time $= \dfrac{5 \cdot 40 + 6 \cdot 15 + 6 \cdot 10 + 5 \cdot 85}{4}$ **(1/2 mark)**

$= 6 \cdot 00$ ms (or $6 \cdot 00 \times 10^{-3}$ s) **(1/2 mark)**

$d = vt$ or $v = \dfrac{d}{t}$ **(1/2 mark)**

$\therefore v = \dfrac{2 \cdot 00}{6 \cdot 00 \times 10^{-3}}$ **(1/2 mark)**

$= 333$ m/s **(1 mark)**

> **HINT** The values in the table are given to 3 significant figures. Therefore, your final answer should also be given to 3 significant figures.

(b)   Move the microphones further apart
or   use a sound source which produces a crisper/sharper/shorter sound, e.g. a 'clapper'.
(1 mark)

> **HINT** A larger distance between the microphones leads to a increased time on the electronic timer. The greater the time measurement, the greater its accuracy.

4.   (a)   (i)   Reverse the bar magnets/magnetic field
or   reverse the direction of the electric current/power supply.   **(1 mark)**

(ii)   Use magnets which create a stronger magnetic field/increase the strength of the magnetic field.   **(1 mark)**

Increase the supply voltage (or current).   **(1 mark)**

(b)   **Carbon (graphite) brushes** – make good electrical contact with the commutator/are good electrical conductors
or   reduce wear (or friction) on the commutator/are a good lubricant/enable the coil to spin freely.   **(1 mark)**

Field coils – create a stronger magnetic field than bar (or permanent) magnets
or   allow the strength of the magnetic field (or speed of motor) to be varied
or   can be used to switch the magnetic field (or the motor) on and off.   **(1 mark)**

> **HINT** You must understand the operation of an electric motor and know the function of the component parts.

5.   (a)

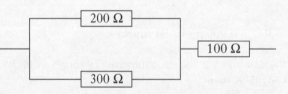

Diagram showing correct resistor combination.   **(1 mark)**

$$\frac{1}{R_T} = \frac{1}{R_1} + \frac{1}{R_2} \quad \textbf{(1/2 mark)}$$

$$\therefore \frac{1}{R_T} = \frac{1}{200} + \frac{1}{300} \quad \textbf{(1/2 mark)}$$

$$\therefore \frac{1}{R_T} = \frac{1}{120}$$

$\therefore R_T = 120 \,\Omega$ **(1/2 mark)** Unit NOT required

$R_T = R_1 + R_2$ **(1/2 mark)**

$\therefore R_T = 120 + 100$ **(1/2 mark)**

$= 220 \,\Omega$ **(1/2 mark)** Unit NOT required

> **HINT** You must be able to calculate the total resistance of a number of resistors (or other circuit components) connected in series or parallel.

(b)

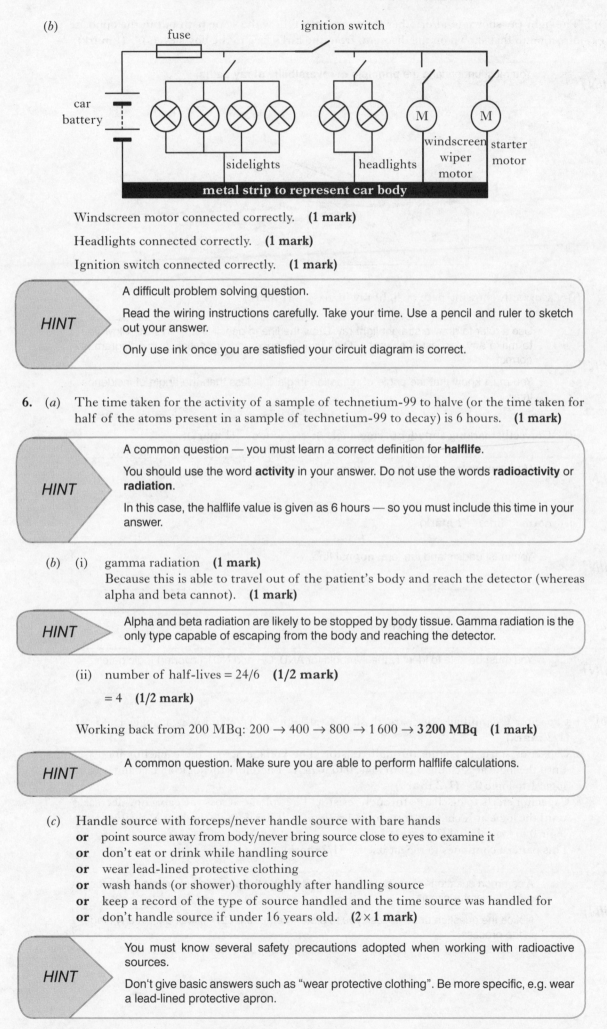

Windscreen motor connected correctly. **(1 mark)**

Headlights connected correctly. **(1 mark)**

Ignition switch connected correctly. **(1 mark)**

> **HINT**
>
> A difficult problem solving question.
>
> Read the wiring instructions carefully. Take your time. Use a pencil and ruler to sketch out your answer.
>
> Only use ink once you are satisfied your circuit diagram is correct.

6. (a) The time taken for the activity of a sample of technetium-99 to halve (or the time taken for half of the atoms present in a sample of technetium-99 to decay) is 6 hours. **(1 mark)**

> **HINT**
>
> A common question — you must learn a correct definition for **halflife**.
>
> You should use the word **activity** in your answer. Do not use the words **radioactivity** or **radiation**.
>
> In this case, the halflife value is given as 6 hours — so you must include this time in your answer.

(b) (i) gamma radiation **(1 mark)**
Because this is able to travel out of the patient's body and reach the detector (whereas alpha and beta cannot). **(1 mark)**

> **HINT**
>
> Alpha and beta radiation are likely to be stopped by body tissue. Gamma radiation is the only type capable of escaping from the body and reaching the detector.

(ii) number of half-lives = 24/6 **(1/2 mark)**

= 4 **(1/2 mark)**

Working back from 200 MBq: 200 → 400 → 800 → 1 600 → **3 200 MBq** **(1 mark)**

> **HINT**
>
> A common question. Make sure you are able to perform halflife calculations.

(c) Handle source with forceps/never handle source with bare hands
or point source away from body/never bring source close to eyes to examine it
or don't eat or drink while handling source
or wear lead-lined protective clothing
or wash hands (or shower) thoroughly after handling source
or keep a record of the type of source handled and the time source was handled for
or don't handle source if under 16 years old. **(2 × 1 mark)**

> **HINT**
>
> You must know several safety precautions adopted when working with radioactive sources.
>
> Don't give basic answers such as "wear protective clothing". Be more specific, e.g. wear a lead-lined protective apron.

7. (a)   The light ray shown (and/or other light rays) can follow the same path but in the opposite direction to that shown on the diagram, from the girl's face to the boy's eye(s).   **(1 mark)**

> **HINT** You must understand the **principle of reversibility of ray paths**.

(b)

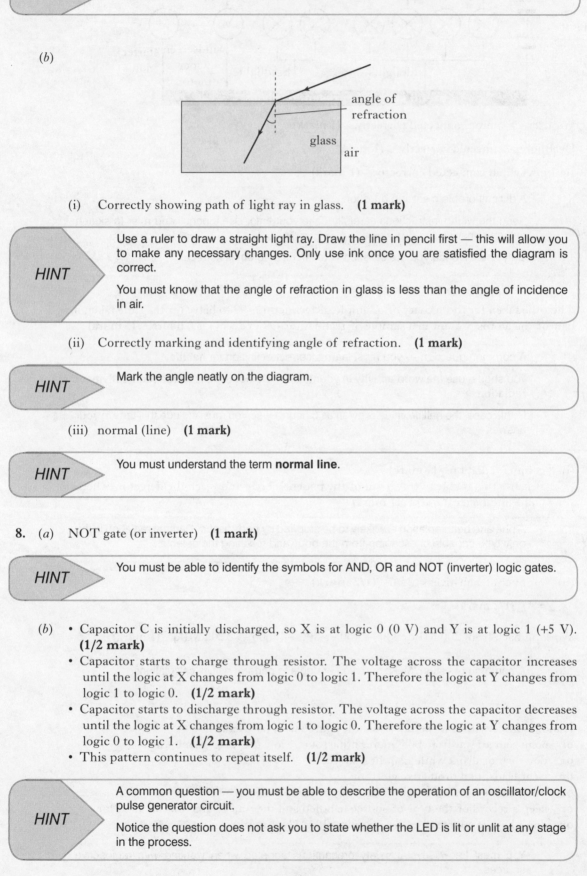

angle of refraction

glass

air

(i)   Correctly showing path of light ray in glass.   **(1 mark)**

> **HINT** Use a ruler to draw a straight light ray. Draw the line in pencil first — this will allow you to make any necessary changes. Only use ink once you are satisfied the diagram is correct.
>
> You must know that the angle of refraction in glass is less than the angle of incidence in air.

(ii)   Correctly marking and identifying angle of refraction.   **(1 mark)**

> **HINT** Mark the angle neatly on the diagram.

(iii)   normal (line)   **(1 mark)**

> **HINT** You must understand the term **normal line**.

8. (a)   NOT gate (or inverter)   **(1 mark)**

> **HINT** You must be able to identify the symbols for AND, OR and NOT (inverter) logic gates.

(b)   • Capacitor C is initially discharged, so X is at logic 0 (0 V) and Y is at logic 1 (+5 V).   **(1/2 mark)**
   • Capacitor starts to charge through resistor. The voltage across the capacitor increases until the logic at X changes from logic 0 to logic 1. Therefore the logic at Y changes from logic 1 to logic 0.   **(1/2 mark)**
   • Capacitor starts to discharge through resistor. The voltage across the capacitor decreases until the logic at X changes from logic 1 to logic 0. Therefore the logic at Y changes from logic 0 to logic 1.   **(1/2 mark)**
   • This pattern continues to repeat itself.   **(1/2 mark)**

> **HINT** A common question — you must be able to describe the operation of an oscillator/clock pulse generator circuit.
>
> Notice the question does not ask you to state whether the LED is lit or unlit at any stage in the process.

(c)   Lower frequency   **(1 mark)**
because capacitor will take a greater time to charge and discharge.   **(1 mark)**

> **HINT**   You must know why increasing the capacitance of the capacitor or increasing the resistance of the resistor gives a longer time between flashes (a lower flash frequency).

(d)   Voltage across R_2 (V_R) $= 5 \cdot 0 - 0 \cdot 5 = 4 \cdot 5$ V   **(1 mark)**

$$V_R = IR_2 \text{ or } R_2 = \frac{V_R}{I} \quad \textbf{(1/2 mark)}$$

$$\therefore R_2 = \frac{4 \cdot 5}{10 \times 10^{-3}} \quad \textbf{(1/2 mark)}$$

$$= 450 \ \Omega \quad \textbf{(1 mark)}$$

> **HINT**   A common question.
>
> This part of the question is worth 3 marks, which indicates more than one part to the calculation.
>
> Many candidates forget to do the first step: calculating the voltage across the resistor. They incorrectly use the voltage across the LED for the second step.

9.   (a) $P = \dfrac{V^2}{R}$ **(1/2 mark)**

$$\therefore P = \frac{12^2}{8 \cdot 0} \quad \textbf{(1/2 mark)}$$

$$\therefore P = 18 \text{ W} \quad \textbf{(1 mark)}$$

(b) $P_{gain} = \dfrac{P_o}{P_i}$ **(1/2 mark)**

$$\therefore P_{gain} = \frac{18}{3 \cdot 0 \times 10^{-3}} \quad \textbf{(1/2 mark)}$$

$$= 6\,000 \quad \textbf{(1 mark)}$$

> **HINT**   Both the top and bottom lines in the power gain relationship must have the same size of unit.
>
> You must change the power input value from 3.0 mW to W — check the multiplication factor on the data sheet.
>
> Power gain, being a ratio, does not have a unit.

(c)   • For one setting of the voltage gain control, read the input voltage across the amplifier from voltmeter V_1 and the output voltage from voltmeter V_2.   **(1/2 mark)**

•   Put the voltage readings obtained into the formula $V_{gain} = \dfrac{V_o}{V_i}$ to calculate the voltage gain.   **(1/2 mark)**

•   Compare the value calculated for the voltage gain with the voltage setting of the voltage gain control.   **(1/2 mark)**

•   Repeat the procedure with different settings of the voltage gain control. **(1/2 mark)**.

> **HINT**   You must know how to measure the voltage gain of an amplifier.

10. (a) (i) $a = \dfrac{v - u}{t}$ **(1/2 mark)**

$$\therefore a = \frac{9 - 0}{5} \quad \textbf{(1/2 mark)}$$

$$= 1 \cdot 8 \text{ m/s}^2 \quad \textbf{(1 mark)}$$

HINT Make sure you read the correct values from the graph.

(ii) The forces acting on the skier are balanced **(1 mark)**
because she is travelling at constant speed. **(1 mark)**

HINT Constant speed indicates balanced forces.

(iii) Decrease in steepness of slope
or moving onto different surface material
or different body position
or increase in friction forces opposing motion. **(1 mark)**

(iv) distance = area under speed-time graph

$$\therefore \text{distance} = (1/2 \times 5 \times 9) + (15 \times 9) \quad \textbf{(1 mark)}$$

$$\therefore \text{distance} = 157 \cdot 5 \text{ m} \quad \textbf{(1 mark)}$$

HINT Make sure you read the correct values from the graph.

Add up the two areas under the graph correctly.

The unit for the distance is **m** not **m²** — a common mistake made by candidates due to the calculation of area.

(b) $E_k = \dfrac{1}{2} mv^2$ or $v = \sqrt{\dfrac{E_k}{1/2 \, m}}$ **(1/2 mark)**

$$\therefore v = \sqrt{\frac{1170}{0 \cdot 5 \times 65}} \quad \textbf{(1/2 mark)}$$

$$= 6 \text{ m/s} \quad \textbf{(1 mark)}$$

HINT The first step is to use the relationship $E_K = 1/2 \, mv^2$ to calculate v^2.

You must then take the square root of this answer to obtain the value for v.

11. (a) unbalanced force (F) = (2 728 + 32) − 200 **(1 mark)**

$$= 2\,560 \text{ N} \quad \textbf{(1 mark)}$$

HINT Take care with your addition and subtraction. Don't mix up the directions of the forces.

(b) $F = ma$ or $a = \dfrac{F}{m}$ **(1/2 mark)**

$$\therefore a = \frac{2\,560}{3 \cdot 2} \quad \textbf{(1/2 mark)}$$

$$= 800 \text{ m/s}^2 \quad \textbf{(1 mark)}$$

HINT You must use the unbalanced force value in your acceleration calculation.

(c)   work done by gravity = decrease in gravitational potential energy

$E_p = mgh$ **(1/2 mark)**

$\therefore E_p = 0.28 \times 10 \times 0.75$ **(1/2 mark)**

$= 2.1$ J   **(1 mark)**

> **HINT**   You must know that, for a falling object:
>
> work done by gravity on object = decrease in gravitational potential energy of object

12. (a) $\dfrac{n_s}{n_p} = \dfrac{v_s}{v_p}$ **(1/2 mark)**

$\therefore \dfrac{110}{n_p} = \dfrac{5}{230}$ **(1/2 mark)**

$\therefore n_p = 5\,060$ turns   **(1 mark)**

> **HINT**   Take care when manipulating the numbers in the transformer relationship.

(b) (i)   Electrical energy is changed to other unwanted forms, heat or sound.   **(1 mark)**
This is due to the transformer coils having resistance
or   due to the creation of small (eddy) currents in the transformer core
or   due to the transformer core being continually magnetised and demagnetised
(hysteresis).   **(1 mark)**

> **HINT**   You must know at least two reasons why transformers are not 100% efficient.

(ii)   percentage efficiency $= \dfrac{\text{useful } P_o}{P_i} \times 100$ **(1/2 mark)**

$\therefore$ percentage efficiency $= \dfrac{I_s V_s}{I_p V_p} \times 100$ **(1/2 mark)**

$= \dfrac{3.45 \times 5}{0.10 \times 230} \times 100$

(1/2 mark for top line and 1/2 mark for bottom line)

$= 75\%$ **(1 mark)**

> **HINT**   Input power is calculated using information provided about the primary coil of the transformer.
>
> Output power is calculated from information provided about the secondary coil of the transformer.

(c)

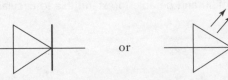

or

(1 mark)

> **HINT**   You must know that diodes and light emitting diodes only allow current to pass through them in one direction.
>
> You should be able to draw the circuit symbol for a diode and light emitting diode (LED).

13. (a) (i)   The liquid wax solidified/turned from a liquid to a solid.   **(1 mark)**

> **HINT**   You must know that when the line on a temperature-time graph is horizontal, the physical state of the substance is changing.

(ii) $E_h = ml$ or $l = \dfrac{E_h}{m}$ **(1/2 mark)**

$\therefore l = \dfrac{40\,500}{0\cdot15}$ **(1/2 mark)**

$= 270\,000$ J/kg (or $2\cdot7 \times 10^5$ J/kg)   **(1 mark)**

> **HINT** You must convert 150 g to kg. (Divide by 1 000). The mass value used in the calculation must be in kg.

(b)   (i)   20°C

> **HINT** Room temperature will be the lowest temperature value shown on the graph. The wax is unable to cool any further.

(ii)   Less time   **(1 mark)**
because copper is a better conductor of heat than glass.   **(1 mark)**

> **HINT** In a given time, more heat energy will travel through the copper beaker from the cooling wax to the surrounding air than through the glass beaker $\therefore$ shorter cooling time with copper.

14. (a)   (i)   If A exerts a force on B, B exerts an equal but opposite force on A.
or   If A pushes B, B pushes A with the same size of force but in the opposite direction.
or   For every action, there is an equal and opposite reaction.   **(1 mark)**

> **HINT** You must know Newton's third law.

(ii)   The exhaust gases pushed out of the rear of the command module by the engine exerted an equal but opposite force on the engine.   **(1 mark)**

> **HINT** You must understand and be able to describe **action and reaction pairs (Newton pairs)**.

(b)   The motion of the command module was a combination of constant horizontal speed and constant downward acceleration due to the attractive force/pull of the moon's gravity. **(1 mark)**
As the command module was pulled towards the moon's surface, the moon's surface curved away from the command module at the same rate, so the command module followed a circular path around the moon.   **(1 mark)**

> **HINT** You must understand projectile motion and be able to extend this to circular motion around a planet or moon.

(c) $W = mg$ or $m = \dfrac{W}{g}$ **(1/2 mark)**

$\therefore$ mass on earth $= \dfrac{104\,760}{10}$ **(1/2 mark)**

$= 10\,476$ kg   **(1 mark)**

Mass on moon is also 10 476 kg because mass remains constant.   **(1 mark)**

> **HINT** You must understand that the mass of an object remains constant, no matter where the mass is in the universe, but that the weight of the object will change as the gravitational field strength acting on the object changes.

15. (*a*) (i) To collect (or focus) light from the star being observed
or to produce an image of the star being observed. **(1 mark)**

(ii) To collect as much light as possible from the star being observed, **(1/2 mark)**
so the image of the star is as bright as possible. **(1/2 mark)**

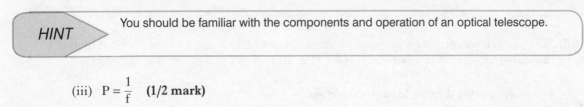

HINT You should be familiar with the components and operation of an optical telescope.

(iii) $P = \dfrac{1}{f}$ **(1/2 mark)**

$\therefore P = \dfrac{1}{500 \times 10^{-3}}$ **(1/2 mark)**

$= (+)\,2\,D$ **(1 mark)**

HINT You must convert the focal length value from mm to m — check the multiplication factor on the data sheet.

The focal length value used in the calculation must be in m.

(*b*)

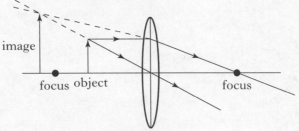

image

focus object   focus

magnifying lens

Drawing light ray from top of object through optical centre of lens. **(1 mark)**

Drawing light ray from top of object, parallel to horizontal axis, to centre line of lens, then through right hand focus. **(1 mark)**

Correct position and size of image. **(1 mark)**

HINT A common question. Make sure you can correctly show how to obtain the size and position of the image.

Light travels in straight lines — you must use a ruler.

Complete the diagram neatly in pencil first — this will allow you to make any necessary changes. Only use ink once you are satisfied that the diagram is correct.

PRACTICE EXAM C   WORKED ANSWERS

1. (*a*) (i) The modulator combines the electrical video signal and the electrical audio signal with the ultra high frequency electrical signal. **(1 mark)**

HINT Use the diagram to help you answer.

Look at the three inputs to the modulator and the single output.

(ii)   The amplifier increases the amplitude/energy of the modulated ultra high frequency electrical signal.   **(1 mark)**

> *HINT*   You must know the purpose of an **amplifier** in an electronic circuit.

(iii)   The modulated ultra high frequency electrical signal is converted to a modulated ultra high frequency television carrier wave.   **(1 mark)**

> *HINT*   Again, use the diagram to help you answer.
>
> Look at the single input to the modulator and the single output.

(b)   The video decoder separates the electrical video signal from the modulated ultra high frequency electrical signal.   **(1 mark)**

> *HINT*   You must know the function of each main part in a radio and television receiver.
>
> Make sure you can draw a **block diagram** for each, correctly labelling all the main parts.

(c)

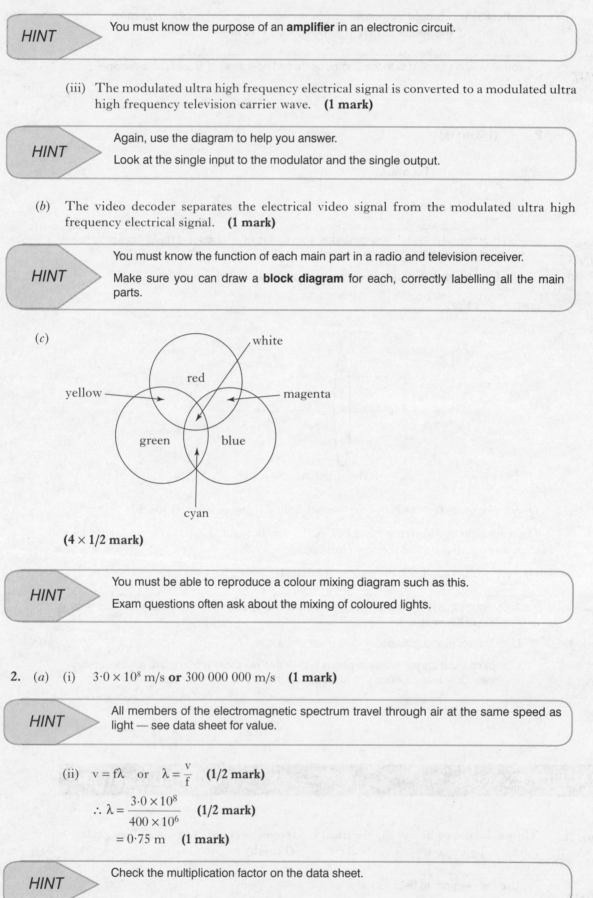

(4 × 1/2 mark)

> *HINT*   You must be able to reproduce a colour mixing diagram such as this.
>
> Exam questions often ask about the mixing of coloured lights.

2. (a) (i) 3.0×10^8 m/s **or** 300 000 000 m/s   **(1 mark)**

> *HINT*   All members of the electromagnetic spectrum travel through air at the same speed as light — see data sheet for value.

(ii) $v = f\lambda$ or $\lambda = \dfrac{v}{f}$ **(1/2 mark)**

$$\therefore \lambda = \frac{3.0 \times 10^8}{400 \times 10^6} \quad \textbf{(1/2 mark)}$$

$$= 0.75 \text{ m} \quad \textbf{(1 mark)}$$

> *HINT*   Check the multiplication factor on the data sheet.

(b) (i) The radio waves are not able to diffract around the buildings
or the radio waves are not able to travel through the buildings. **(1 mark)**

(ii) long wave **or** low frequency **(1 mark)**

HINT
Make sure you understand **diffraction** — the ability of waves to travel around solid obstacles.

Long wave (low frequency) waves diffract more than short wave (high frequency) waves.

3. (a) electrical energy to light energy. **(1 mark)**

HINT
You must know that electrical signals travel through metal wires and light signals travel through optical fibres.

(b) Lower cost
or lighter weight
or higher signal capacity/carry more signals
or better signal quality/less interference
or less decrease in signal strength/less repeaters (or amplifiers) required. **2 × 1 mark**

HINT
Optical fibres have several advantages over metal wires for telecommunication purposes. The only advantages metal wires have over optical fibres are:

• higher signal speed in metal wires;

• metal wires are easier to join.

(c) (i)

Either correct position of **i**. **(1 mark)**

Correct position of **r**. **(1 mark)**

HINT
Write the letters neatly in the correct position.

(ii) The angle of incidence between ray B and the normal line is less than the critical angle for the optical fibre/is too small for total internal reflection to occur. **(1 mark)**

HINT
You must know the condition necessary for total internal reflection to take place.

4. (a)

Voltage across resistor R (V)	Current through resistor R (A)	Voltage across resistor R / Current through resistor R
6·0	0·030	200
7·0	0·035	200
8·0	0·040	200
9·0	0·045	200

All three values correct. **(1 mark)**

> **HINT** This part of the question involves three simple calculations, hence its 1 mark value.

(b) Quantity remains constant/stays the same/does not change. **(1 mark)**

> **HINT** The answer should be obvious from the three values you calculated in part (a).

(c) By changing/adjusting the resistance of the variable resistor. **(1 mark)**

> **HINT** You must know that decreasing the resistance of the variable resistor increases the size of the voltage across resistor R and the size of the current passing through it.

(d) $200\ \Omega$ **(1 mark)**

> **HINT** This is the ratio you calculated in part (a).

(e) $P = I^2R$ **(1/2 mark)**

$\therefore P = 0{\cdot}030^2 \times 200$ **(1/2 mark)**

$= 0{\cdot}18\ \text{W}$ **(1/2 mark)**

> **HINT** A common question.
>
> Don't forget to square the current value.

5. (a) $1\ \text{kWh} = 1\ \text{kW} \times 1\ \text{hour}$ **(1/2 mark)**

$= 1\,000\ \text{W} \times 3\,600\ \text{s}$ **(1/2 mark)**

$= 1\,000\ \text{J/s} \times 3\,600\ \text{s}$ **(1/2 mark)**

$= 3\,600\,000\ \text{J}$ **(1/2 mark)**

> **HINT** You must know that 1 hour = 3 600 s and that 1 W = 1 J/s.

(b) (i) Total power consumption $= 100 + 60 + 40 + 2\,000 = 2\,200\ \text{W}$ **(1/2 mark)**

$= 2{\cdot}2\ \text{kW}$ **(1/2 mark)**

> **HINT** Remember to change W to KW.
>
> Check the multiplication factor on the data sheet.

(ii) $E = Pt$ **(1/2 mark)**

$\therefore E = 2{\cdot}2\ \text{kW} \times 2\ \text{h}$ **(1/2 mark)**

$\mathbf{E} = 4{\cdot}4\ \text{kWh}$ **(1/2 mark)**

$\text{Cost} = 4{\cdot}4\ \text{kWh} \times 15\ \text{p} = 66\ \text{p}$ **(1/2 mark)**

> **HINT** You must know how to perform cost calculations involving kWh.

6. (*a*) (i)

retina

The three light rays meeting at the back of the eye. **(1 mark)**

> **HINT** You must be able to draw the path of light rays after they have passed through convex and concave lenses.
>
> Use a ruler to draw straight light rays. Draw the rays in pencil first — this will allow you to make any necessary changes. Only use ink once you are satisfied that the diagram is correct.

    (ii) The light rays are parallel. **(1 mark)**

> **HINT** You must know that light rays from a distant object are parallel.

    (iii)

Thicker lens. **(1 mark)**

> **HINT** When focusing on close objects, the thickness of the eye lens increases.

(*b*) (i) $P = \dfrac{1}{f}$ or $f = \dfrac{1}{p}$ **(1/2 mark)**

        $\therefore f = \dfrac{1}{-2 \cdot 5}$ **(1/2 mark)**

        $= -0 \cdot 4$ m **(1/2 mark)**

> **HINT** The negative (–) sign must be included in all calculation steps and the final answer.

    (ii) short-sight-edmess **(1 mark)**

> **HINT** (ii) Concave lenses have a negative (–) power and are used to correct short sight.
>
> Convex lenses have a positive (+) power and are used to correct long sight.

7. (*a*) (i) Ultrasound is sound (or vibrations) with frequency above the upper range of human hearing/greater than 20 000 Hz. **(1 mark)**

> **HINT** You must learn a correct definition of **ultrasound**.

(ii) • Pulses of ultrasound are passed through the body from a transducer placed on the skin. **(1/2 mark)**

• The pulses of ultrasound reflect off the kidneys/kidney stones, back to the transducer. **(1/2 mark)**

• The transducer detects the strength of the reflected pulses and passes this information to a computer. **(1/2 mark)**

• The computer builds up an image of the kidney/kidney stones. **(1/2 mark)**

> **HINT** A common question that can be applied to many body parts, e.g. scan of heart or unborn baby in womb. This technique is also used to detect cancer tumours.
>
> Describe each step clearly. Use the word "reflect". Do not use vague terms such as "bounce off".

(iii) 1500 m/s **(1 mark)**

> **HINT** See data sheet for value, and make sure you select the value for **tissue**.

(iv) $d = vt$ or $t = \dfrac{d}{v}$ **(1/2 mark)**

$\therefore t = \dfrac{0 \cdot 03}{1\,500}$ **(1/2 mark)**

$= 2 \times 10^{-5}$ s **(1 mark)**

> **HINT** You must convert 3 cm to m. (Divide by 100). The distance value used in the calculation must be in m.

(b) (i) photographic film **(1 mark)**

> **HINT** You must know a suitable detector for every member of the electromagnetic spectrum.

(ii) Prolonged exposure to X-rays can damage (or kill) body cells/organs. **(1 mark)**

> **HINT** You must know that overexposure to X-rays can harm body tissue.

(iii) The X-rays are always directed at the tumour. **(1 mark)**
The X-rays pass through different parts of the surrounding tissue, so cause much less damage to the surrounding tissue. **(1 mark)**

> **HINT** Think carefully about the situation presented.
>
> This is a 2 mark question, so your answer should include 2 reasons.

8. (a) (i) OR gate **(1 mark)**
(ii) AND gate **(1 mark)**

> **HINT** You must be able to identify the truth tables for AND, OR and NOT (inverter) logic gates.

(*b*) (i)

A	B	C	D	E	F	G
0	0	0	1	1	0	0
0	0	1	1	1	0	0
0	1	0	0	0	1	0
0	1	1	0	1	0	0
1	0	0	1	1	0	0
1	0	1	1	1	0	0
1	1	0	0	0	1	1
1	1	1	0	1	0	0

All entries in column E correct. **(1 mark)**

All entries in column F correct. **(1 mark)**

All entries in column G correct. **(1 mark)**

HINT Complete the truth table in pencil first — this will allow you to make any necessary changes. Only use ink once you are satisfied all the entries in the table are correct.

(ii) When switched on, the iron bar in the centre of a solenoid is attracted into the solenoid — thus allowing the padlock to open. **(1 mark)**

HINT You must understand how a solenoid operates.

9. (*a*) (i) Decrease the resistance of the variable resistor **(1 mark)**
and decrease the capacitance of the capacitor. **(1 mark)**

HINT A common question. You must be able to describe the operation of an oscillator/clock pulse generator circuit.

You must know that decreasing the resistance of the resistor or decreasing the capacitance of the capacitor increases the frequency of the clock pulses.

Be specific. Do not use terms such as "use a smaller resistor or capacitor" — you will not be awarded any marks for such an answer.

(ii) $1001 = (1 \times 8) + (0 \times 4) + (0 \times 2) + (1 \times 1)$ **(1 mark)**

$\qquad = 8 + 0 + 0 + 1$ **(1/2 mark)**

$\qquad = 9$ **(1/2 mark)**

HINT You must know how to convert a binary number to a decimal number.

(*b*) $P_{gain} = \dfrac{P_o}{P_i}$ or $P_o = P_{gain} \times P_i$ **(1/2 mark)**

$\therefore P_o = 750 \times 3{\cdot}6$ **(1/2 mark)**

$\qquad = 2\,700 \text{ mW}$ (or $2{\cdot}7$ W) **(1 mark)**

HINT It is acceptable to use **mW** as the unit for your answer.

Be warned — if you try to convert your answer to W and you make an error, you will lose marks.

DO NOT PERFORM ADDITIONAL CALCULATIONS WHICH ARE NOT NECESSARY.

10. (*a*) (i) Decrease in height = 4 000 – 1 200 **(1/2 mark)**

= 2 800 m **(1/2 mark)**

E_p = mgh **(1/2 mark)**

∴ E_p = 72·0 × 10 × 2 800 **(1/2 mark)**

E = 2 016 000 J (or 2·016 × 10⁶ J) **(1 mark)**

> **HINT** Rather than perform two gravitational potential energy questions, it is preferable to calculate the **decrease in height**, then use this value in the relationship E_p = mgh.

(ii) decrease in gravitational potential energy = increase in kinetic energy **(1/2 mark)**

∴ 2 016 000 = 1/2 mv² **(1/2 mark)**

∴ $v^2 = \dfrac{2\,016\,000}{1/2\ m}$

$= \dfrac{2\,016\,000}{0\cdot5 \times 72}$ **(1/2 mark)**

= 56 000 **(1/2 mark)**

∴ v = 237 m/s **(1 mark)**

> **HINT** A statement about using an appropriate number of significant figures in the final answer to a question is given on the front cover of the exam paper.
>
> In this case, 3 significant figures are sufficient.

(*b*) (i) Point R **(1 mark)**

> **HINT** There is a large decrease in downward speed at the instant the parachute is opened.

(ii) PQ **(1/2 mark)**
and RS **(1/2 mark)**

> **HINT** When the forces acting on an object are unbalanced, the speed of the object changes.

(*c*) Fall head (or feet) first
or move arms/legs closer to body
or curl into ball shape. **(1 mark)**

> **HINT** You must know ways to reduce or increase the size of the frictional forces acting on objects.

---

11. (*a*) (i) d = vt or $v = \dfrac{d}{t} = \dfrac{\text{length of mask}}{\text{time for mask to pass through light gate}}$ **(1/2 mark)**

∴ $v = \dfrac{0\cdot06}{0\cdot08}$ **(1/2 mark)**

= 0·75 m/s **(1 mark)**

> **HINT** Make sure you use the **time taken for the mask to pass through the light gate** in your calculation.

(ii) $a = \dfrac{v - u}{t}$ **(1/2 mark)**

$= \dfrac{0\cdot75 - 0}{0\cdot50}$ **(1/2 mark)**

$= 1\cdot5 \text{ m/s}^2$ **(1 mark)**

HINT   Make sure you read the correct values from the table.

(b)   (i)   Human reaction time makes the timing of short time intervals using a stopwatch inaccurate.   **(1 mark)**

HINT   You must know that **human reaction time** makes the use of a stopwatch or stopclock for measuring short time intervals inaccurate.

(c)   decreased   **(1 mark)**
A shorter length of mask will take less time to pass through the light gate.   **(1 mark)**

HINT   You must know that the shorter the time interval measured, the closer the calculated speed of a moving object is to its instantaneous speed.

12.   (a)   (i)   Rotating the bar magnet creates a changing magnetic field next to the coil of copper wire.   **(1 mark)**

HINT   You must know that a voltage is induced (created) across a conductor when the conductor is stationary in a changing magnetic field or the conductor is moved in a constant magnetic field.

(ii)

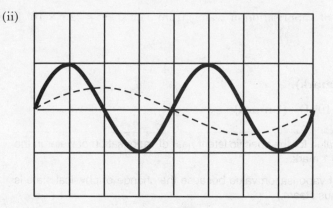

As shown by dashed line, trace should have lower amplitude   **(1 mark)**
and   should be spread further apart.   **(1 mark)**

HINT   Lower speed of rotation gives lower voltage (lower trace amplitude) and lower frequency (less wavelengths on trace).

Do not be tricked by the extra space provided above and below the trace — the amplitude of the trace decreases.

(iii)   increased voltage   **(1 mark)**

HINT   You must know three ways of increasing the size of the voltage induced across a conductor.

(*b*) (i) In a full size a.c. generator:

- there are field coils/electromagnets instead of the bar magnet;
- the field coils/electromagnets rotate;
- there is more than one coil of wire;
- the coils of wire are stationary;
- the coils of wire contain more turns of wire. **(2 × 1 mark)**

HINT You must know these differences.

(ii) A. Wind turbines do not create atmospheric pollution/acid rain **(1 mark)**

B. Backup sources of electrical energy are required for when the amount of electrical energy produced by wind turbines is low. **(1 mark)**

HINT You must know the advantages and disadvantages of several renewable energy sources.

13. (*a*) Value of 'c' (specific heat capacity of water) from data sheet = 4 180 J/kg °C **(1 mark)**

$E_h = mc\Delta T$ **(1/2 mark)**

$\therefore E_h = 4\,180 \times 1{\cdot}5 \times (100 - 22)$ **(1/2 mark)**

$\therefore E_h = 489\,060$ J **(1 mark)**

HINT You need to look up the value for the specific heat capacity of water in the data sheet — this is worth 1 mark.

Be careful. Use the temperature **difference** in your $E_h = mc\Delta T$ calculation.

(*b*) Value of 'l' (specific latent heat of vaporisation of water) from data sheet = $22{\cdot}6 \times 10^5$ J/kg **(1 mark)**

$E_h = ml$ **(1/2 mark)**

$\therefore E_h = 1{\cdot}5 \times (22{\cdot}6 \times 10^5)$ **(1/2 mark)**

$\therefore E_h = 3\,390\,000$ J (or $3{\cdot}39 \times 10^6$ J) **(1 mark)**

HINT You need to look up the value for the specific latent heat of vaporisation of water in the data sheet — this is worth 1 mark.

(We use the latent heat of vaporisation value because the change of physical state is from liquid water to gaseous steam.)

(*c*) $P = \dfrac{E}{t}$ or $t = \dfrac{E}{P}$ **(1/2 mark)**

$\therefore t = \dfrac{3\,390\,000}{2\,000}$ **(1/2 mark)**

= 1695 s **(1 mark)**

HINT The steamer is only useful while steam is being produced — not while the water is heating up. Therefore, the heat energy value obtained from part (b) of the question should be used in the power calculation.

(d) (i) $P = IV$ or $I = \dfrac{P}{V}$ **(1/2 mark)**

$\therefore I = \dfrac{2\,000}{230}$ **(1/2 mark)**

$= 8\cdot7$ A to 2 significant figures **(1 mark)**

> **HINT**
>
> The answer to your calculation should be given to 2 significant figures.

(ii) $Q = It$ **(1/2 mark)**

$\therefore Q = 8\cdot7 \times (25 \times 60)$ **(1/2 mark)**

$= 13\,050$ C **(1 mark)**

> **HINT**
>
> Make sure you convert 25 minutes to seconds correctly. (Multiply by 60).
>
> The time value used in the calculation must be in s.

14. (a) Gravitational field strength is the ratio of weight to mass for an object
or the weight per unit mass of an object
or the force (of gravity) acting on every kg of an object. **(1 mark)**

> **HINT**
>
> A common question. You must learn a correct definition for **gravitational field strength**.

(b) (i) 6 N/kg **(1 mark)**

(ii) $W = mg$ **(1/2 mark)**

$\therefore W = 2\,600 \times 6$ **(1/2 mark)**

$= 15\,600$ N **(1 mark)**

> **HINT**
>
> Make sure you use the value for the gravitational field strength at a height of 1750 km above the surface of the earth in your calculation.

15. (a) X = gamma rays **(1 mark)**

Y = microwaves **(1 mark)**

> **HINT**
>
> You must know all the members of the electromagnetic spectrum and be able to arrange them in order of increasing or decreasing wavelength or frequency.

(b) If 1 wavelength (1λ) is produced in time (T), a wave will travel a distance (d) of 1 wavelength (1λ) in time (T):

$$\text{speed (v)} = \frac{\text{distance (d)}}{\text{time (T)}} = \frac{1\,\lambda}{T} = \frac{1}{T} \times \lambda \quad \textbf{(1 mark)}$$

$= f \times \lambda$ **(1 mark)**

since $f = \dfrac{1}{T}$

> **HINT**
>
> Many candidates have difficulty with this question.
>
> You must learn to explain why the two relationships are equivalent.

1. (*a*) (i) So the television signals do not interfere with one another. **(1 mark)**

(ii) Channel 4 **(1 mark)**

 HINT Highest frequency is equivalent to shortest wavelength — if a source produces more waves every second, the wave fronts are closer together.

(*b*)

curved reflector

transmitter

- Rays from transmitter hit curved reflector. **(1/2 mark)**
- Reflected rays are straight — ruler used. **(1/2 mark)**
- Reflected rays are parallel. **(1/2 mark)**
- Arrow heads present to show correct direction. **(1/2 mark)**

HINT Use a ruler to draw the signals as straight lines. Draw the lines in pencil first — this will allow you to make any necessary changes. Only use ink once you are satisfied the diagram is correct.

This part of the question is about **transmitted** signals — make sure the direction of your arrow heads is **away from** the transmitter.

(*c*) **Line build-up** — individual pictures on the television screen are built up from separate lines which are 'traced' onto the screen by a moving electron beam. **(1 mark)**

Image retention — the human eye/brain system holds an image for a fraction of a second before replacing it with another image. It blends successive pictures on the television screen together, so we observe a moving picture. **(1 mark)**

Brightness variation — phosphor material on the inside surface of the television screen glows for a fraction of a second every time it is hit by an electron from the electron beam. Brighter parts of a picture are created by more electrons in the electron beam hitting those parts of the television screen. **(1 mark)**

HINT You must know how a single picture is built up on the screen of a television and why there is the illusion of continuous movement on the screen.

You must not mention modern LCD or plasma screen televisions in your answer.

2. (*a*) (i) The spy satellite orbits the earth several times every day. **(1 mark)**
At the same time, the earth spins beneath it, so different areas of the earth's surface can be observed. **(1 mark)**

HINT Think carefully. Use information provided in the question.

(ii)   A. 300 000 000 m/s (or 3.0×10^8 m/s)   **(1 mark)**

> *HINT*   All members of the electromagnetic spectrum travel through air at the same speed as light — see data sheet

B. $v = f\lambda$ or $f = \dfrac{v}{\lambda}$ **(1/2 mark)**

$\therefore f = \dfrac{3.0 \times 10^8}{0.10}$ **(1/2 mark)**

$\quad = 3.0 \times 10^9$ Hz   **(1 mark)**

> *HINT*   You must convert the wavelength value from cm to m. (Divide by 100).

C. $d = vt$ or $t = \dfrac{d}{v}$ **(1/2 mark)**

$\therefore t = \dfrac{750 \times 10^3}{3.0 \times 10^8}$ **(1/2 mark)**

$\quad = 2.5 \times 10^{-3}$ s (or 0.0025 s)   **(1 mark)**

> *HINT*   You must convert the distance value from km to m — check the multiplication factor on the data sheet.
>
> The distance value used in the calculation must be in m.

(b)   The telecommunication satellite stays above the same point on the earth's surface. **(1 mark)**

It is too high to obtain close-up images of the earth's surface.   **(1 mark)**

3.   (a)   Peak voltage is higher than the value usually quoted.   **(1 mark)**

> *HINT*   You must be able to compare the peak value of an a.c. supply with the quoted value. The quoted value is always less than the peak value.

(b)   (i) $Q = It$ or $t = \dfrac{Q}{I}$ **(1/2 mark)**

$\therefore t = \dfrac{6.0}{4.8 \times 10^{-3}}$ **(1/2 mark)**

$\quad = 125$ s   **(1 mark)**

> *HINT*   You must convert the current value from mA to A — check the multiplication factor on the data sheet.
>
> The current value used in the calculation must be in A.

(ii)   Charges will receive a greater quantity of energy.   **(1 mark)**

> *HINT*   Higher voltage means higher energy.

(iii) A. **Brushes** — carry electric current between the battery/power supply and the commutator. **(1 mark)**

Commutator — reverses the direction of the electric current flowing in the coil every time the coil makes a half turn. **(1 mark)**

> *HINT* You must understand the operation of an electric motor and know the function of the component parts.

B. The electric current in one side of the coil is in the opposite direction to the current in the other side of the coil. **(1 mark)**

4. (a) (i) $R_T = R_1 + R_2 + R_3$ **(1/2 mark)**

$\therefore R_T = 7.2 + 7.2 + 7.2$ **(1/2 mark)**

$\therefore R_T = 21.6 \ \Omega$ **(1 mark)**

> *HINT* You must be able to calculate the total resistance of a number of resistors (or other circuit components) connected in series or parallel.
>
> In this circuit, the lamps are connected to the ohmmeter in a continuous loop, so are in series.

(ii) If one lamp was faulty, the others could not light

or the lamps could not be switched on and off separately

or the lamps could not light brightly. **(1 mark)**

> *HINT* You must know why household lighting circuits are connected in parallel, not series.

(b) (i)

point in circuit	P	Q	R	S
size of electric current at point in circuit (A)	3·4	1·7	1·7	3·4

(4 × 1/2 mark) (Marks will be awarded for values of Q and R which add up to 3·4).

> *HINT* On entering a ring main circuit, the current splits up — half usually travels around the circuit in a clockwise direction while the other half travels anticlockwise.

(ii) a.c. (or alternating current) supply **(1 mark)**

> *HINT* Current from an alternating current (a.c.) supply constantly reverses direction at a set time interval.

(iii) Ring main circuit uses thinner/less expensive cable

or ring main cable carries lower current

or ring main cable does not create as much heat. **(1 mark)**

> *HINT* You must know two advantages a ring circuit has over a standard parallel circuit for connecting appliances in parallel.

(c)　P = IV

$$\therefore P = I \times (IR) \quad \textbf{(1 mark)}$$

$$\therefore P = I^2R \quad \textbf{(1 mark)}$$

> **HINT**
>
> Many candidates have difficulty with this question.
>
> You must learn how to obtain the relationship P = I²R from P = IV and V = IR.

5. (a) (i)　total internal reflection　**(1 mark)**

> **HINT**
>
> You must know that light travels along an optical fibre by **total internal reflection**.

　　(ii)　To carry light reflected from inside the patient's body to
　　　　　the surgeon's eye.　**(1 mark)**

> **HINT**
>
> You must be able to explain how an endoscope works.

(b) (i)　Laser beam has high energy so can vaporise/burn away tumour.
　　　　or Heat from laser beam will seal blood vessels/'bloodless surgery'.　**(1 mark)**

> **HINT**
>
> You must know that laser light carries a large amount of energy.

　　(ii)　• Locate tumour using cold light from endoscope.　**(1/2 mark)**

　　　　• Fix position of endoscope.　**(1/2 mark)**

　　　　• Direct laser beam onto tumour by passing beam down optical fibre bundle X.
　　　　　(1/2 mark)

　　　　• Use cold light from endoscope to observe whether tumour has been vaporised.
　　　　　(1/2 mark)

> **HINT**
>
> Think carefully about your answer. You should be able to explain how an endoscope works and how it can be used for medical surgery.
>
> Describe each step clearly.

(c)　Computerised tomography provides a more detailed/three-dimensional image
　　or shows the exact position and size of the tumour.　**(1 mark)**

> **HINT**
>
> You must know the advantages computerised tomography (CT) scans have over common X-rays.

6. (a) (i)　Gamma radiation is able to pass out of the body. (Alpha radiation can't.)　**(1 mark)**
　　　　　Gamma radiation does not cause great damage to the body. (Alpha radiation does.)
　　　　　(1 mark)

> **HINT**
>
> You must know that alpha radiation ionises atoms and therefore damages/kills body cells.
>
> You must also know that alpha radiation has a short range, so will not be able to pass out of the body.

(ii)   right kidney.   **(1 mark)**
This is taking a much longer time than the left kidney to pass the radioactive liquid tracer to the patient's bladder.   **(1 mark)**

> **HINT**   Think carefully. Use the information provided at the start of the question and from the graph.

(iii)   tracer C   **(1 mark)**
The activity of the radioactive tracer while in the body will not decrease significantly during the investigation process but will decrease to a low/safer level in a reasonable time afterwards.   **(1 mark)**

> **HINT**   You must understand how the halflife value of a radioactive substance can affect its use.

(b)   (i)   • The (type of) absorbing tissue.   **(1 mark)**
            • The nature/type of radiation.   **(1 mark)**

> **HINT**   You must know the two factors upon which the biological effect of radiation depends.

(ii)   sievert   **(1 mark)**

> **HINT**   You must know the unit for dose equivalent.
> Do not give the unit symbol (Sv) — The mark will not be awarded for the symbol alone.

(c)   **Geiger-Müller tube** — radiation entering the tube ionises gas atoms/molecules inside the tube.   **(1 mark)**
This produces electrons/electrical pulses which are counted.   **(1 mark)**
or
Film badge — radiation striking photographic film affects chemicals on the surface of the film.   **(1 mark)**
When the film is developed, the amount of darkening/fogging on the image gives an indication of the radiation exposure.   **(1 mark)**
or
Scintillation counter — radiation entering the counter is absorbed by a fluorescent substance.   **(1 mark)**
This converts the energy of the radiation into tiny pulses of light which are counted.   **(1 mark)**

> **HINT**   You must be able to describe how radiation is detected by Geiger-Muller tubes, film badges and scintillation counters.

7.   (a)   (i)   A '1' output is required when the lock is broken. Logic gate X changes a '0' output from the contact switch to a '1'.   **(1 mark)**

> **HINT**   Think carefully. Examine the circuit design and read the logic instruction for the contact switch in the lock.

(ii)

INPUT A	INPUT C	OUTPUT D
0	0	0
0	1	1
1	0	1
1	1	1

- Correct headings. **(1/2 mark)**
- Correct pairs of input values for A and C. **(1 mark)**
- Correct output value D for each pair of input values A and C. **(1/2 mark)**

> **HINT**
>
> You must be able to identify the symbols for AND, OR and NOT (inverter) logic gates. You must also be able to draw the truth table for each of these logic gates when no input values are provided.
>
> Complete the truth table in pencil first — This will allow you to make any necessary changes. Only use ink once you are satisfied all the entries in the table are correct.

(b) Voltage across R (V_R) = $12 - 2 \cdot 2 = 9 \cdot 8$ V **(1 mark)**

$$V_R = IR \text{ or } R = \frac{V_R}{I} \quad \textbf{(1/2 mark)}$$

$$\therefore R = \frac{9 \cdot 8}{10 \times 10^{-3}} \quad \textbf{(1/2 mark)}$$

$$= 980 \ \Omega \quad \textbf{(1 mark)}$$

> **HINT**
>
> A common question.
>
> This part of the question is worth 3 marks, which indicates more than one part to the calculation.
>
> Many candidates forget to do the first step: calculating the voltage across the resistor. They incorrectly use the voltage across the LED for the second step.

8. (a) $V_{ldr} = \dfrac{R_{ldr}}{R_{ldr} + R_{var\ res}} \times V_S$ **(1/2 mark)**

$$\therefore V_{ldr} = \frac{18}{18 + 2} \times 6 \quad \textbf{(1/2 mark)}$$

$$= 5 \cdot 4 \text{ V} \quad \textbf{(1 mark)}$$

> **HINT**
>
> This question involves a voltage divider. Write the resistance values on the circuit diagram so you can see the correct order in which to write the values in the voltage divider relationship.

(b) $V_{var\ res} = V_{supply} - V_{ldr}$

$$\therefore V_{var\ res} = 6 \cdot 0 - 5 \cdot 4 \quad \textbf{(1/2 mark)}$$

$$= 0 \cdot 6 \text{ V} \quad \textbf{(1/2 mark)}$$

> **HINT**
>
> This part of the question is worth 1 mark — so it is unlikely you will have to use a relationship from the data book.
>
> Remember that, in a voltage divider, the voltage across each resistor adds up to the supply voltage.

(c)   Increase resistance of variable resistor.   **(1 mark)**

> **HINT**
>
> You must know how altering the resistance of the different components in the voltage divider part of such a transistor switching circuit affects whether the transistor will switch the output device on or off.

9.   (a)   distance = area under speed-time graph

∴ distance $= (1/2 \times 2{\cdot}5 \times 12) + (1{\cdot}5 \times 12) + (1/2 \times 2 \times 12)$ **(1 mark)**

$= 45$ m   **(1 mark)**

> **HINT**
>
> Make sure you read the correct values from the graph.
>
> Split the area under the graph into two triangles and one rectangle, then calculate the area of each shape using values from the graph axes. Do not use the number of boxes! Add up the three areas under the graph correctly to obtain the total distance travelled.
>
> The unit for the distance is **m** not **m²** — a common mistake made by candidates due to the calculation of area.

(b)   (i) $d = vt$ or $v = \dfrac{d}{t}$ **(1/2 mark)**

∴ $v = \dfrac{4{\cdot}5}{6}$ **(1/2 mark)**

$= 7{\cdot}5$ m/s   **(1 mark)**

> **HINT**
>
> Make sure you divide the total distance travelled by the total time taken.

(ii)   During a journey, the instantaneous speed of an object keeps changing.   **(1 mark)**

(c)   (i) $a = \dfrac{v - u}{t}$ **(1/2 mark)**

∴ $a = \dfrac{0 - 12}{2}$ **(1/2 mark)**

$= -6$ m/s²   **(1 mark)**

> **HINT**
>
> You must be able to identify the downward slope on the graph (from 4 to 6 s) as deceleration.
>
> Take care. Make sure you read the correct values from the graph.
>
> Because this is a deceleration, the answer to your calculation must have a negative (–) value.

(ii) $F = ma$ **(1/2 mark)**

∴ $F = 68 \times 6$ **(1/2 mark)**

$= 408$ N   **(1 mark)**

> **HINT**
>
> There is no need to include the negative sign for the acceleration in your calculation. If you do include the negative sign, you must include a negative sign in front of your calculated force value.

10. (*a*) Due to 'human reaction time'
or
there is a time delay between seeing an event and reacting to it. **(1 mark)**

> **HINT** — You must know that **human reaction time** makes it impossible to react to an incident/event at the instant it happens.

(*b*) d = vt **(1/2 mark)**

Higher speed x constant reaction time = greater distance travelled
(1/2 mark)

> **HINT** — This is a difficult question. Increased speed multiplied by a constant reaction time gives a larger distance.

(*c*) Brakes have to do more work (changing increased kinetic energy to heat energy).
(1 mark)
Braking force is constant, so longer time is required. **(1 mark)**

> **HINT** — You must know that the greater the kinetic energy of a moving object, the greater the amount of work required to stop the object.

11. (*a*) Neutrons released hit more uranium nuclei. **(1 mark)**
This releases a greater number of neutrons which continue the process. **(1 mark)**.

> **HINT** — You must know how **chain reactions** are started and why they continue.

(*b*) Increase the number of turns of wire in the coils
or increase the strength of the magnetic field
or increase the speed of rotation of coils. **(2 × 1 mark)**

> **HINT** — You must know three ways of increasing the size of the voltage induced across a conductor.

(*c*) (i) $\dfrac{n_s}{n_p} = \dfrac{V_s}{V_p}$ **(1/2 mark)**

$\therefore \dfrac{n_s}{1\,000\,000} = \dfrac{400}{25}$ **(1/2 mark)**

$\therefore n_s = 16\,000\,000$ turns (or $1 \cdot 6 \times 10^7$) turns **(1 mark)**

> **HINT** — Take care when manipulating the numbers in the transformer relationship.

(ii) Some electrical energy is changed to other unwanted forms of energy such as heat or sound. **(1 mark)**

> **HINT** — You must know at least two reasons why transformers are not 100% efficient.

(d)　Total resistance of power line = 75 km × 0·36 Ω/km　**(1/2 mark)**

　　　 = 27 Ω　**(1/2 mark)**

　　　 ∴ P = I²R　**(1/2 mark)**

　　　 ∴ P = (5 × 10³)² × 27　**(1/2 mark)**

　　　 ∴ P = 6·75 × 10⁸ W (or 675 000 000 W)　**(1 mark)**

> **HINT**　This part of the question is worth 3 marks, which indicates more than one part to the calculation.

12. (a)　$P = \dfrac{E}{t}$ or E = Pt　**(1/2 mark)**

　　　　 ∴ E = 100 × 5　**(1/2 mark)**

　　　　 　= 500 J　**(1 mark)**

(b)　E_p = mgh　**(1/2 mark)**

　　　 ∴ E_p = 2·5 × 10 × 3·2　**(1/2 mark)**

　　　 　= 80 J　**(1 mark)**

(c)　percentage efficiency = $\dfrac{\text{useful } P_o}{P_i} \times 100$　**(1/2 mark)**

　　　 ∴ percentage efficiency = $\dfrac{80}{500} \times 100$　**(1/2 mark)**

　　　　　　 = 16 %　**(1 mark)**

>
> **HINT**　The energy input is the energy supplied by the electric motor.
>
> The energy output is the gravitational potential energy gained by the mass.
>
> Make sure you put these values in the correct place in the percentage efficiency relationship. Remember, it is impossible to have an efficiency greater than 100%!

13. (a)　(i)　W = mg　**(1/2 mark)**

　　　　　　 ∴ W = (1·5 × 10⁶) × 10　**(1/2 mark)**

　　　　　　　　 = (1·5 × 10⁷) N or 15 000 000 N　**(1 mark)**

> **HINT**　Remember to use the earth's gravitational field strength value in your calculation — see data sheet for value.

　　　(ii)　Mass decreases due to fuel being used up
　　　　　 or gravitational field strength decreases as rocket travels further away from earth.
　　　　　 (1 mark)

> **HINT**　As a rocket travels, it burns fuel, so the mass of the rocket decreases, so its weight will also decrease.
>
> As a rocket travels further away from the earth's surface, the gravitational field strength of the earth acting on the rocket decreases, so the weight of the rocket also decreases.

(b)

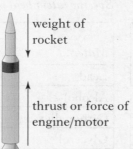

weight of rocket

thrust or force of engine/motor

Each correctly identified force.   **(1/2 mark)**

Each correct direction.   **(1/2 mark)**

> **HINT**   Use clear arrows to show the direction of each force.

14. (a) $E_k = \frac{1}{2}mv^2$ **(1/2 mark)**

E_k before entering atmosphere $= 1/2 \times 120 \times 2\,000^2$ **(1/2 mark)**

$= 240\,000\,000$ J   **(1/2 mark)**

E_k after entering atmosphere $= 1/2 \times 120 \times 900^2$

$= 48\,600\,000$ J   **(1/2 mark)**

$\therefore$ decrease in $E_k = 240\,000\,000 - 48\,600\,000$ **(1/2 mark)**

$= 191\,400\,000$ J

> **HINT**   This part of the question is worth 3 marks, which indicates more than one part to the calculation.
>
> Take care — each separate step involves large numbers.

(b) decrease in kinetic energy = increase in heat energy   **(1 mark)**

$\therefore 191\,400\,000 = cm\Delta T$ **(1/2 mark)**

$\therefore T = \dfrac{191\,400\,000}{cm}$

$\therefore T = \dfrac{191\,400\,000}{1\,100 \times 120}$ **(1/2 mark)**

$= 1\,450\,°C$ **(1 mark)**

> **HINT**   **'Rise in temperature'** suggests use of the relationship $E_h = mc\,\Delta T$.

(c) Heat energy may escape to the atmosphere

or   heat energy may melt part of meteorite.   **(1 mark)**

> **HINT**   You must know that heat energy can be lost to the surroundings and that when heat energy is used to melt a substance, the temperature of the substance (while melting) does not change.

DATA SHEET FOR CREDIT PHYSICS PRACTICE EXAM PAPERS

Gravitational field strengths

	Gravitational field strength on the surface in N/kg
Sun	270
Mercury	4
Venus	9
Earth	10
Moon	1·6
Mars	4
Jupiter	26
Saturn	11
Neptune	12

Specific latent heat of fusion of materials

Material	Specific latent heat of fusion in J/kg
Lead	$0·25 \times 10^5$
Alcohol	$0·99 \times 10^5$
Carbon dioxide	$1·80 \times 10^5$
Glycerol	$1·81 \times 10^5$
Copper	$2·05 \times 10^5$
Water	$3·34 \times 10^5$
Aluminium	$3·95 \times 10^5$

Melting and boiling points of materials

Material	Melting point in °C	Boiling point in °C
Alcohol	−98	65
Turpentine	−10	156
Glycerol	18	290
Lead	328	1737
Aluminium	660	2470
Copper	1077	2567

Specific latent heat of vaporisation of materials

Material	Specific latent heat of vaporaisation in J/kg
Turpentine	$2·90 \times 10^5$
Carbon dioxide	$3·77 \times 10^5$
Glycerol	$8·30 \times 10^5$
Alcohol	$11·2 \times 10^5$
Water	$22·6 \times 10^5$

SI Prefixes and Multiplication Factors

Prefix	Symbol	Factor	
nano	n	0.000 000 001	$= 10^{-9}$
micro	μ	0.000 001	$= 10^{-6}$
milli	m	0.0001	$= 10^{-3}$
kilo	k	1000	$= 10^3$
mega	M	1 000 000	$= 10^6$
giga	G	1 000 000 000	$= 10^9$

Speed of light in materials

Material	Speed in m/s
Diamond	$1·2 \times 10^8$
Glass	$2·0 \times 10^8$
Glycerol	$2·1 \times 10^8$
Water	$2·3 \times 10^8$
Air	$3·0 \times 10^8$
Carbon dioxide	$3·0 \times 10^8$

Specific heat capacity of materials

Material	Specific heat capacity in J/kg °C
Lead	128
Copper	386
Glass	500
Aluminium	902
Silica	1033
Ice	2100
Glycerol	2400
Alcohol	2350
Water	4180

Speed of sound in materials

Material	Speed in m/s
Carbon dioxide	270
Air	340
Tissue	1500
Water	1500
Muscle	1600
Glycerol	1900
Bone	4100
Aluminium	5200
Steel	5200